Data centres: an introduction to concepts and design

CIBSE Knowledge Series: KS18

Authors
Jim Harrison, Paul Hood, David Hughes,
Guy Hutchins, Colin Hyde, Phil Jolly, Mick Marshall,
George Moss, Owen O'Connell, Mark Seymour,
Alex Sharp, Arthur Spirling, Robert Thorogood,
Robert Tozer, Brian Waddell

Contributors
Derek Allen, Christian Beier, Alan Beresford,
Sophia Flucker, Leon Markwell, Ender Ozbeck,
Mike Smith, Kevin Stanley

Editor
Bonnie Craig

CIBSE Head of Knowledge
Nicholas Peake

CIBSE

Contents

1   Introduction

This document provides an introduction to owners, co-location developers, designers, constructors, operators and all those who have an interest in data centre design, operation and space planning. It gives an introduction to many of the concepts that require careful consideration. This document is not intended to be a design tool, but should be used for guidance only as to the more significant issues that might be considered.

Over recent years data centres have gained greater significance and complexity in the way they are designed and engineered. They are not necessarily a commercial enterprise within themselves but more of a significant and necessary component in the way businesses perform and operate.

Customers range from the most demanding, where downtime and failures cannot be tolerated, through to the commercial, industrial and scientific communities. For example, the banking and financial sector cannot tolerate system failures or accept prolonged periods of downtime. They therefore require highly engineered solutions that, by inference, are of a complex nature although within manageable proportions. The more risk-averse business users, such as financiers and gaming institutions, can be closely regulated. This might drive their behaviour and use of space, resulting in high resilience and conservative design. Other end users accept lower levels of resilience providing it is properly detailed and managed. This might include a level of structured downtime with more frequent service intervals and perhaps no system backup or strategic, off-site, backup facility.

The size of an information technology (IT) or data centre installation can vary enormously from a few kW for the small commercial user to those serving large financial institutions of many mW with server rack densities of 4 kW not being unusual. This could relate to approximately 2 kW/m^2 for a fully populated, high-density server facility. Although this range represents many magnitudes of scale, the engineering solutions in many ways remain the same.

With owning and operating a data-processing facility comes the responsibility of minimising operating costs, including energy consumption and usage. This requires specialist knowledge and expertise, especially when considering energy consumption, flexibility of operation, legacy installations and planning for the future.

Figure 1:
**Indicative data centre
components**
PDU: power distribution unit
CRAC: computer room air conditioning
UPS: uninterrupted power supply

2   The function of a data centre

In broad terms a data centre typically consists of the components in Figure 1.

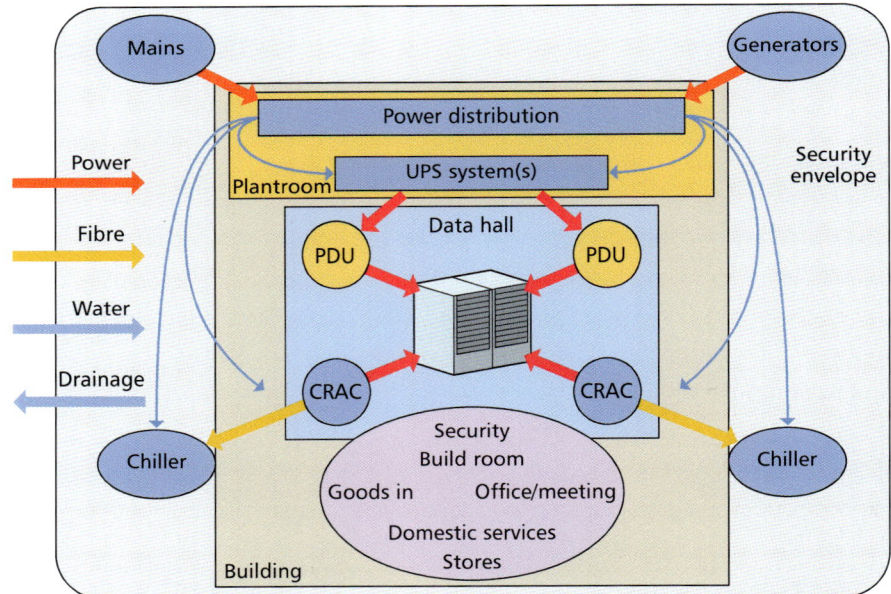

Intent: *To provide a secure and conditioned environment
for electronic and data processing equipment*

Most computer equipment requires an environment with specific ranges of temperature, humidity and cleanliness requirements. High-quality power must be delivered to the equipment within acceptable ranges of voltage and frequency transients. The criticality of the information being processed requires the facility to be physically secure and therefore generally incorporates surveillance and access control.

Access to high-quality, readily available, highly resilient IT is now the lifeblood of the vast majority of organisations. Any interruption to, or loss of, the vital data processing function could seriously disrupt the organisation's ability to carry out, or in a worst case carry on, its business. Although the 'cloud' (cloud computing is the delivery of computing as a service rather than a product) promises much in terms of such data processing, many organisations will continue to run their own data centres and the associated hardware and network infrastructure. It is essential therefore that the design and sizing of both the primary and secondary data centres, if used, be appropriate for current needs and the inevitable expansion of the data processing function. Due consideration should also be given to disaster recovery plans and/or support facilities.

The data centre eventually houses a variety of hardware from a range of vendors in varieties of cabinets. These will range from low-density cabinets, e.g. housing network or computing equipment consuming 1 or 2 kW of power to more modern high density cabinets, e.g. full of blade processors consuming 5 to 10 kW to high performance computing cabinets consuming 25 to 35 kW. In addition, there will be high-density file and backup systems containing the invaluable data on which the whole enterprise depends. Cabinets from some vendors may be designed to be water cooled or refrigerant cooled. A cable and fibre network infrastructure will serve the data centre cabinets and reach to the rest of the organisation and to external network providers to provide internet access. Figure 2 shows a bird's eye view of a data centre.

Consideration must be given to the likely growth in IT demand in both the increase in transactional processing but also the need to store and backup more and more active data. For some industries there is an increasing need to store archive data for longer periods to meet compliance regulations.

IT equipment has a typical life cycle of two to five years (although it may need to interact with systems that are much older) with the building and engineering plant having a much longer life expectancy. To avoid early obsolescence, the design needs to allow for future power density (increases) and different IT configurations.

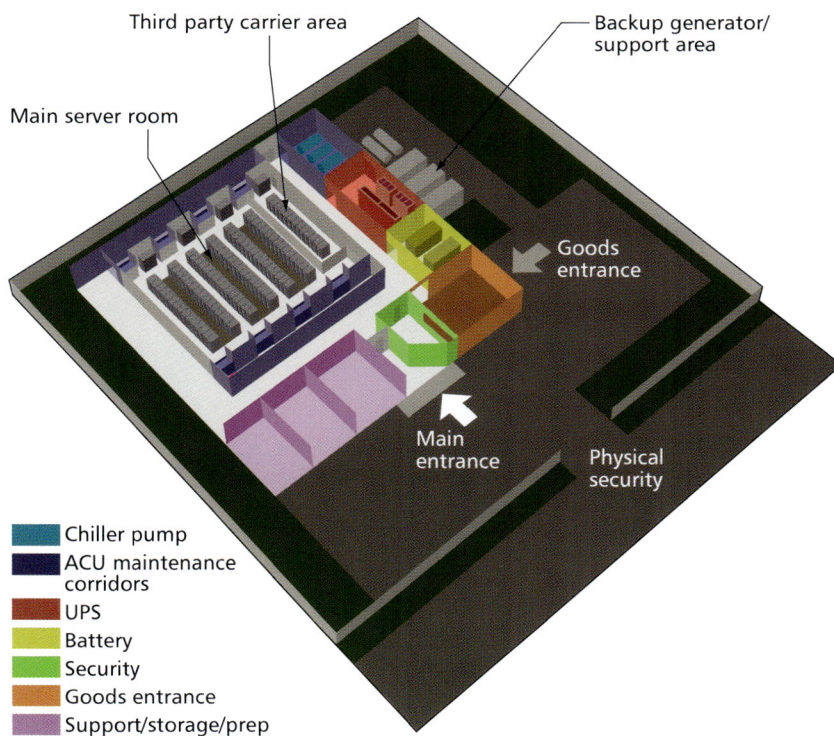

Third party carrier area
Backup generator/support area
Main server room
Goods entrance
Main entrance
Physical security

Chiller pump
ACU maintenance corridors
UPS
Battery
Security
Goods entrance
Support/storage/prep

Figure 2:
Bird's eye view of a data centre
ACU: air conditioning unit

The data centre design should also consider changes in IT, e.g. server and storage virtualisation, which are leading to increased utilisation and hence power demand per cabinet. Higher component packing densities—both for servers and disc arrays—are also leading to increased power consumption per m^2 and to an increase in weight loading. These modern technologies are more tolerant of higher operating temperatures. This is recognised in modern data centre design and helps to reduce power consumption. However it should be remembered that some legacy IT equipment, which would be very expensive to replace due to the criticality of the software running on it, cannot run at these higher temperatures but will still need to be catered for. Designers need to agree with operators how older, less tolerant IT platforms can be accommodated if required.

All of the equipment needs to be provided with the appropriate power, cooling, fire suppression, security and networking capability with a level of resilience relevant to the organisation's operation. Those operating a highly transactional service 24/7 will need a higher level of resilience than those organisations less dependent on IT for their day-to-day business. All data centre designs should reflect the risk profile of the business that they serve.

3   Site selection and building criteria

3.1   Introduction

This section sets out a basis for evaluating potential sites when considering a location for a new build data centre. The location for a data centre is influenced by a large number of factors including how it will be run and be operated. Will it be operated as a 'dark' site with few attendant personnel apart from security in attendance or are transportation links important?

The brief should clearly detail what the site needs to provide for the operator. The appointment of experienced advisors and consultants will refine this so a specification can be produced to assist in the evaluation and validation for potential sites. Climate may play an important role in site selection from the point of view of free-cooling and cost-operating expenditure and should be evaluated as part of the initial analysis.

The overall due diligence of the sites visited by the team should include a risk analysis. This will assist in producing a meaningful report and should include a 'gap analysis' in relation to the client brief.

Planning both site and internal space is an important part of the design process in interpreting and delivering a project that meets the client's aspirations. This is explained in more detail in section 4.

3.2   Key considerations

— Appointment of a professional team of advisors.

— Compilation of the brief including items such as resilience levels and tier rating (see appendix A) as these will have a large impact on planning consent, space and cost.

— Availability of utility services for power, water, data and telecom links.

— Day 1 and day 2 expansion provision for growth.

— Clear energy and sustainability measures put in place; these are major factors that will affect accreditation schemes covered by: Building Research Establishment Environmental Assessment Method (BREEAM); Leadership in Energy and Environmental Design (LEED); Green Star.

— Power density normally expressed in W/m^2 with an indication of individual peak cabinet loads extrapolated from the client's initial and perceived future requirements. Also the segregation of low- and high-density racks to afford uniform rack cooling and the avoidance of 'hot spots'.

— Will the facility be standalone or a paired data centre? (Distances between sites will impact on communication speeds.)

Typically, data halls would be subdivided into low-/high-diversity areas. Table 1 provides suggested guidance on both cabinet and power density. The table is based on average performance across a number of installations, although it is accepted that in some situations these figures can be exceeded.

Table 1:

Classification guide of rack loads and popular cooling systems

Description	Heat load per rack (kW)	Power density (W/m²)	Typical cooling medium	Cooling system
Low density	1–7	500–900	Air	CRAC or CRAH units All-air systems
Medium density	8–10 10–14	900–1500	Air CW/refrigerant/ carbon dioxide	Hot or cold aisle containment Containment and in-row liquid cooling
High density	15–24	5000+	CW/refrigerant/ carbon dioxide	Cabinet/rear door liquid cooling
High density plus+	25+	8000+	CW/refrigerant/ carbon dioxide	Cabinet/rear door liquid cooling
CRAC: computer room air conditioning; CRAH: computer room air handling; CW: chilled water				

The relative importance of each of these criteria will depend on the nature of the business being served.

3.3 Building design criteria

— Location and transportation links.

— Power and fibre: capacity, resilience, diversity and connectivity.

— Location relative to latency and/or asynchronous replication and fibre distance.

— Interconnectivity to external systems, such as wide area network (WAN), local area network (LAN).

— Planning: building classification and use.

— Building control regulations and approval.

— Energy and sustainability.

— Landscape.

— Climate.

Building criteria

— Single or multi-storey facility.

- Structural grid, floor loadings, building envelop and foundation design.

- Slab-to-slab heights, wall types.

- Lift/elevator (capacity) and access routes for equipment (impacts on corridor and door dimensions), e.g. plant replacement/maintenance strategies.

- Raised floor design (load bearing capacity), depth and access ramps.

- Vehicular access, deliveries.

- Disability Discrimination Act (DDA) compliance, disability access and compliance with local codes.

3.4 Risk analysis

Approach to risk review

The attitude towards how risk is understood, reviewed and accepted for a data centre is extremely important. No two organisations or companies are the same in how they consider risk. For data centres, it is important that the employing organisation has a good understanding of its own operating risks and what impact risk or particular events would have on their business. The following should therefore be taken into account:

- determination of the overall governance of a project

- the overall governance of the ongoing facility.

The following risks should be considered.

External site risks:

- Terrorist or other security threats.

- Aircraft, airports and airfields.

- Security exclusion zones.

- Climate/environment risks.

- Environmental impact risk to others.

- Geotechnical/seismic risk.

- Potential risk to availability of utilities at some future date (power, fibre, gas, water, drainage).

- Acoustic constraints.

- Future power/energy strategy and sources.

Internal site/design risks:

— Internal configuration of data halls, internal plant and support areas.

— Water leaks through the building fabric.

— External plant and equipment and the impact of vibration.

— Man/machine interfaces.

— Facilities management (FM) and appropriate levels of technical training.

— Ease of access, protection of fabric.

— Ability to provide future expansion capability.

— Ability to comply with target corporate social responsibility (CSR) plus others such as BREEAM, LEED and Green Star in Australia.

— Establish requirements for business continuity.

— Reliability and redundancy.

— Facilities management and maintenance.

— Risk of system instability, both in steady state under dynamic conditions and low operation.

— Working in an existing 'live data centre' environment.

— Security: hardened facilities, internal and external access control, electro-magnetic interference (EMI), closed circuit television (CCTV) and lighting.

— Fire strategy and brigade access.

— Emergency provisions (fire, blast, terrorist, contamination etc).

— Health and safety (H&S) risks during the project and thereafter.

— Security risk and strategy (terrorism/intruder/natural and man-made disasters).

— Project delivery risk associated with unclear brief, late changes, unrealistic budgets or unrealistic programmes.

All of the above should be brought together in a risk schedule, compiled with input from all project stake holders, analysed and assessed, leading to a plan of work in order to mitigate risk to an acceptable level. This is an ongoing process and as the design phases progress, these risks should be designed out through the project and effectively handed over to the operational team on completion.

3.5   Planning risks

—   Archaeology and heritage.

—   Change of planning classification/use.

—   Control of major accident hazards covered by local codes.

—   Fuel storage.

—   Ecology.

—   Local objections.

—   Noise.

—   Any carbon tax or renewables contribution.

The above criteria and characteristics must satisfy the client's cost and programme requirements.

Typically, timeframes for the delivery of these facilities do not differ between refurbished and new-build, but are dependent on the size and will intrinsically be linked to the delivery of power and fibre (which can be lengthy and dependent on upgrade of existing or new supplies).

4   Planning, site and space

4.1   Site planning

Planning constraints are numerous and vary from country to country. Full detailed planning in the UK will take between 8 and 13 weeks depending on the complexity of the development. A pre-application including an initial consultation with the planning department is highly recommended as it enables the professional team to identify key areas of local authority concern. The intention during this period is to formalise a workshop between the professional team and representations of the various statutory bodies to accurately produce the level of detailed reports, studies, surveys and project risk prediction required. The normal response time in the UK on a pre-application is three to four weeks. These will cover several issues that are normally highlighted by the submission and will include, but will not be limited to, the following points of reference:

—   noise from a major plant in relation to the ambient background level normally based on the early hours of the morning (this may limit the ability to maximise the use of free cooling at night)

—   traffic, car parking, travel plan

—   drainage

—   rights of way and local access codes

—   landscaping

—   ecology

—   visual impact

—   archaeology and heritage

—   environmental impact assessment

—   sustainability

—   building materials

—   change of use

—   control of major accident hazard, covered by legislation.

The team will be looking for an early indication from the planning department of the suitability of the proposal in terms of massing, siting and general impact. Any previous planning history on the chosen site will also help to identify local issues and areas of potential conflict (probably not an issue on very small sites). In general the pre-application should decrease the determination period.

During the determination period, the UK statutory consultation period is 28 days, during which the local authority engages with all concerned parties, i.e. local residents and internal departments. In line with current statutory legislation, the application, when submitted, will be accessible through planning portals so regular checks can be made with relevant actions taken to minimise the scale of conditions likely to be placed on any decision notice.

4.2   Space planning

Key strategic points that the internal layout of both the data halls and plant areas are listed here.

Level of resilience required to support the IT equipment:

—   basic (N)

—   redundant components (N+1)

—   concurrently maintainable 2(N)

—   fault tolerant (2N) sometimes known as (S+S) system plus system.

The Uptime Institute (http://uptimeinstitute.com) and document TIA 942 (TIA, 2012) provide further guidance on this.

Varying levels of resilience also impose greater demands on space as a result of diverse or non-diverse cable and pipe routes.

Equipment to be installed within the data halls:

—   cabinet sizes

—   density kW/m^2

—   means of fire escape and travel distance

—   legacy and new equipment (legacy equipment refers to older or existing IT equipment, which may be retained and may require closer environmental control than modern equipment).

Figure 3 shows the key elements and adjacencies for a typical data centre.

Figure 3:
Key elements and adjacencies for a typical data centre

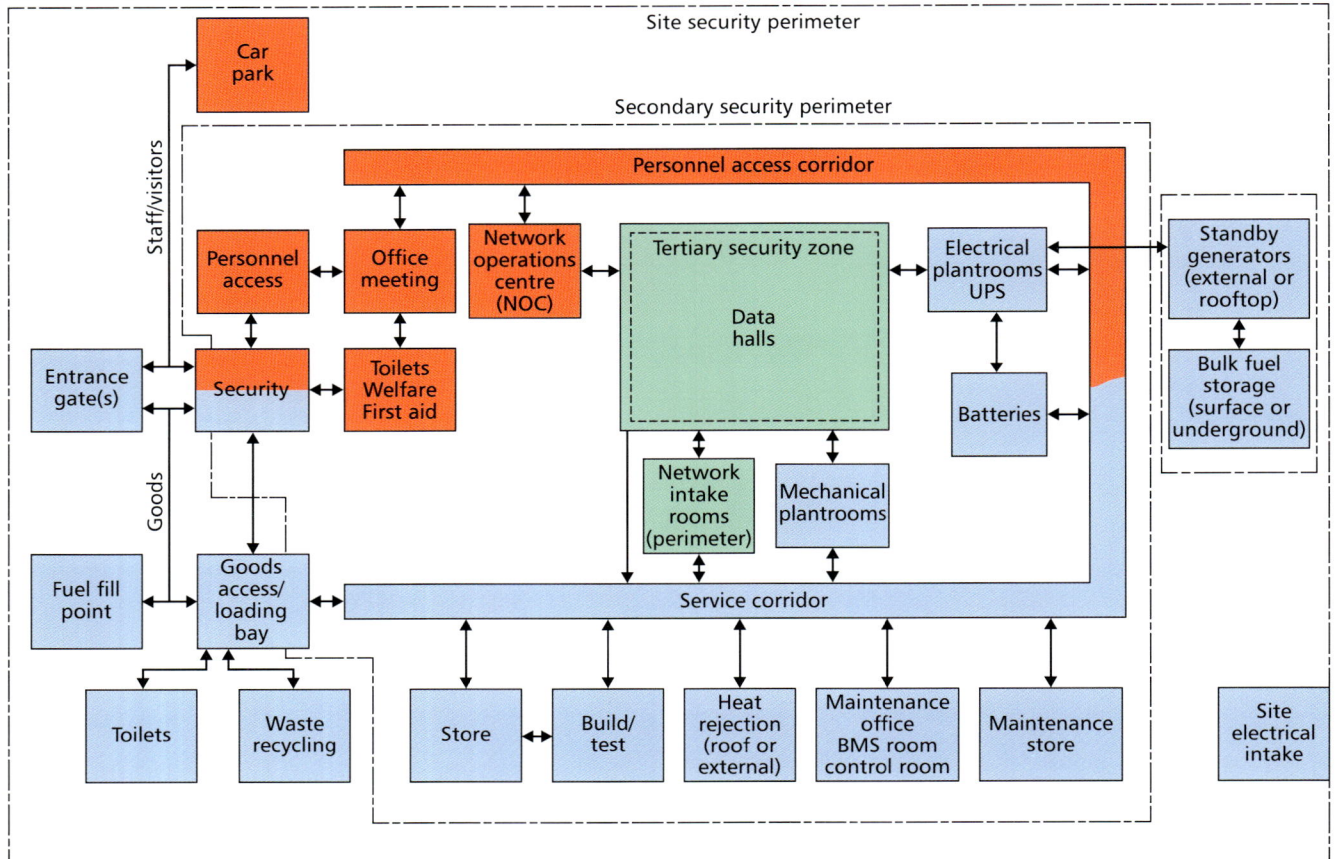

BMS = building management system

5   Energy and sustainability

Rising energy costs, together with ever-increasing demand for IT services and equipment, have led to increased business interest in operating costs, including data-centre energy use and efficiency.

This, coupled with a continuing focus on carbon reduction in many countries, will require companies to ensure their data centre energy consumption is reduced to meet local targets.

There is a heavy focus on energy efficiency in the data-centre industry, where a lot of useful information is produced and exchanged by such bodies as The Green Grid and the American Society for Heating, Refrigeration and Air Conditioning Engineers (ASHRAE) Technical Committee 9.9. There is general consensus that the air entering IT equipment can be as high as 27 °C (or even higher, although these temperatures may not be appropriate for legacy equipment), however consideration should be given to the room 'rate of rise' in a failure scenario.

With this higher server rack supply air temperature, consideration must be given to maximising any available free cooling minimising the use of mechanical refrigeration.

This provides an opportunity for free cooling, with countries such as the UK having temperate climates with temperatures suitably low for significant periods of the year. It is important to consider the IT server rack supply temperature and humidity requirements and to understand whether there is a possibility to operate at higher temperatures/wider humidity range and realise energy savings on the data-centre cooling system.

5.1   Energy strategy

Energy efficiency can be targeted at each layer of the data centre infrastructure starting from the IT equipment itself (see Figure 4).

The proposed energy strategy should allow for this interdependence as follows.

IT equipment

—   Increased utilisation of IT equipment.

—   Rationalise processing.

Figure 4:
Data-centre energy strategy

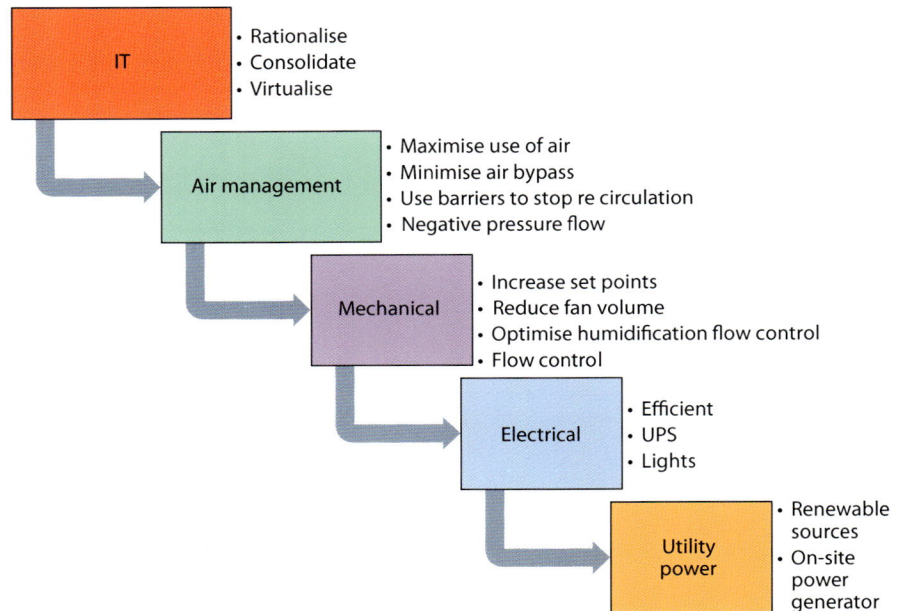

IT	• Rationalise • Consolidate • Virtualise
Air management	• Maximise use of air • Minimise air bypass • Use barriers to stop re circulation • Negative pressure flow
Mechanical	• Increase set points • Reduce fan volume • Optimise humidification flow control • Flow control
Electrical	• Efficient • UPS • Lights
Utility power	• Renewable sources • On-site power generator

— More efficient IT equipment (increased processing output per unit of energy input) and correct size processing.

— Higher range of inlet temperatures above current standards where and how temperatures are measured and their frequency of measurement.

— Ensure IT equipment is correctly selected and not over specified. Beware of specified peak loads and loads operating somewhere less than peak load (part of requirements capture).

Air management

Determining the cooling strategy and its management is important. Using cool air effectively is a key enabler of mechanical systems energy saving. Minimising bypass air/air leakage allows the reduction of fan energy consumption. Minimising hot air exhausted by IT equipment recirculating into IT equipment inlets allows set points from local computer room air conditioning (CRAC)/ computer room air handling (CRAH) units to be raised, where refrigeration equipment works at higher coefficient of performance (COP) and cooling systems can operate with more hours of free cooling (when and if available).

Mechanical

Establishing an efficient cooling strategy is a key factor and using psychometric charts overlaid with the energy relationship helps clarify this. The set points of the cooling cycle have the biggest effect on the efficiency of refrigeration systems. Increasing these above accepted norms is often one of the largest energy savers of the facility, due to their impact on the cooling cycle and improved opportunities for free cooling. If free cooling can be achieved all year round there is no need for mechanical refrigeration systems. Air management is an important enabler for achieving energy efficiency objectives through reduced refrigeration and CRAC/CRAH fan energy requirements.

Electrical

These involve many items of which the uninterruptible power supply (UPS) is normally the predominant loss:

—   UPS: increase part load/efficiency and operate energy-saving features

—   luminaires: use more efficient luminaires and automatic operating systems

—   generator pre-heaters operate to required temperatures (not more)

—   other electrical losses: normally involve eliminating redundant plant, increasing part loads and improving quality of energy supply, i.e. less harmonic content, etc.

—   the use of variable speed drives/motors particularly at low load conditions.

Utility power

Once the core energy-saving measures have been addressed, improving the energy supply options to the site may be considered. The strategic principle to consider is that the carbon dioxide emissions produced by the alternative source, e.g. cogeneration, are lower over the site's lifetime than the equivalent produced by the mains grid.

Consider how each country has been and will be improving the efficiency of its power sources. Furthermore, if renewable bio-fuel energy is considered, e.g. rapeseed oil, due attention should be paid to its sustainability, i.e. will there be enough arable land to meet the world's demands in the future? The following points should be considered, amongst others:

—   whether a base thermal load exists cogeneration with lower carbon dioxide emissions than the national grid over the site's lifetime, e.g. using sustainable bio-fuel

—   photo-voltaic solar cells (although impact is small and consideration needs to be given regarding additional structural costs for supporting large arrays)

—   wind turbines

—   purchasing renewable energy from utility (usually at a cost premium)

—   possibly other distribution network operators (DNO).

5.2 Efficiency of IT equipment

IT technology is driving toward higher utilisation and densities, which is resulting in higher IT equipment air discharge temperatures. Blade servers can have discharge air temperatures up to 25 °C higher than the air inlet at full load. Therefore, with higher densities and utilisation, hot air stream temperatures can be expected to be in the range of 50 °C and higher. This provides a further opportunity for partial free cooling because cooling outdoor air will be at a lower temperature for most parts of the year than cooling return air from the IT equipment. Care should also be taken to observe health and safety requirements when adopting air containment systems with high temperatures.

5.3 Power usage effectiveness

How effectively power is used can dramatically affect energy consumption and carbon emissions. One measure that has been adopted by the industry is known as the power usage effectiveness (PUE). This is the ratio of data-centre energy to IT equipment measurement, defined by The Green Grid as an annual average value. Typical values for legacy facilities vary between 2.0 and 2.5. Current design focuses on PUE values of less than 1.4 although the trend is to be even lower. A PUE value of 2 means that twice as much power is being used in the data centre and its infrastructure as a whole than is required by the IT itself. As all the services load factors are based on the IT equipment loads, purchasing better, more efficient IT equipment means the whole site becomes more energy efficient with a consequent carbon reduction.

It should be remembered that PUE can fluctuate depending on IT load and only reaches its design value when a data centre becomes fully populated. Figure 5 shows that for a legacy data centre, the single largest consumer

Figure 5:
Typical energy uses: legacy (e.g. PUE = 2.0)

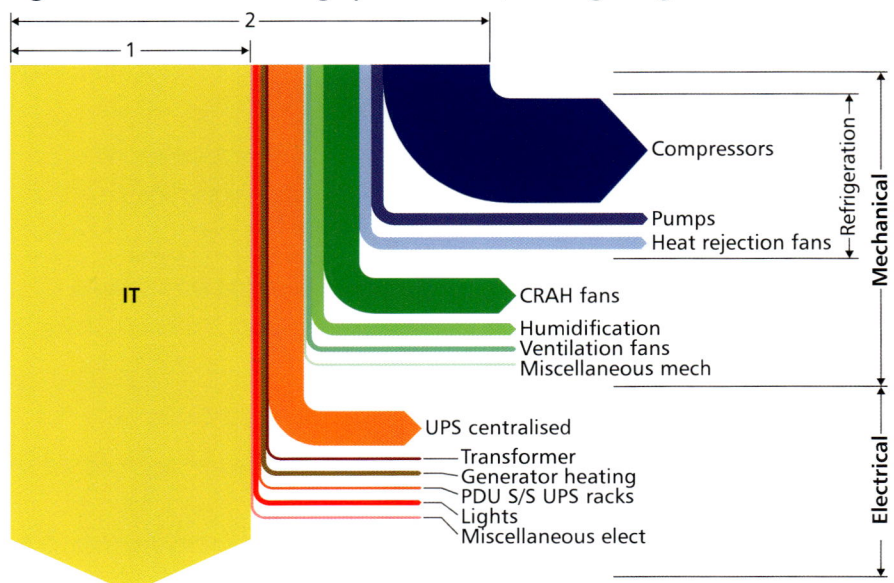

of power is compressor power whereas in current design compressor power can often be reduced by free cooling (see Figure 6 for typical energy uses in current designs).

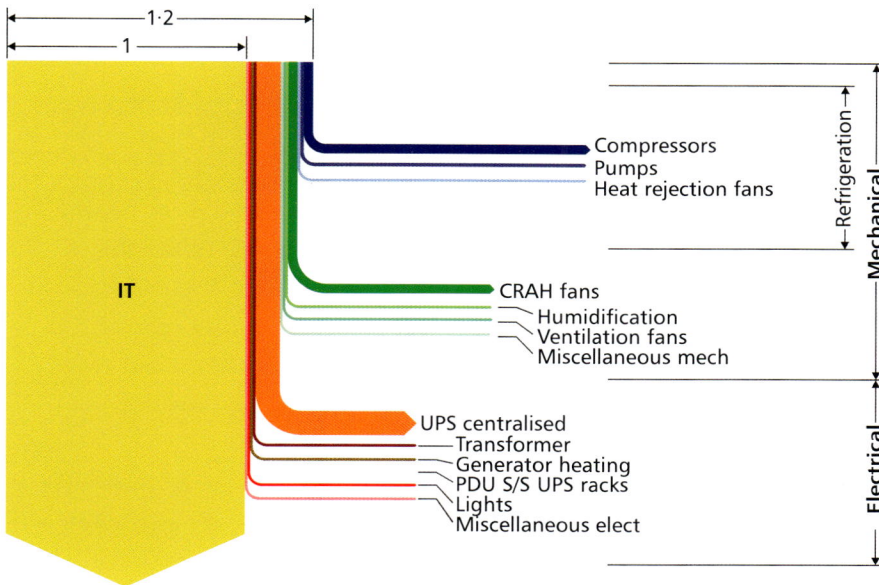

5.4   Environmental benchmarking

European Code of Conduct

In 2008 the European Commission launched a voluntary Code of Conduct for data-centre energy efficiency (EU, 2008), with the aim of promoting best practice and reducing energy consumption (known as EU CoC 2008). Data-centre operators can apply to become participants and commit to reporting their energy consumption and implementation of a series of best practices across their data centre, which cover the following areas:

— data centre utilisation, management and planning

— IT equipment and services

— cooling

— data-centre power equipment

— other data-centre equipment

— data-centre building.

Co-location operators and managed service providers have a reduced scope according to their areas of responsibility, compared with enterprise data-centre operators.

Other data-centre stakeholders, such as vendors, consultancies (design, engineering, maintenance and service companies), utilities, customers of data-centre services, industry associations/standards bodies and educational institutions, can apply to become endorsers of the code and show that they are working to support data centre operators in meeting the objectives of the Code of Conduct. The code is publicly available and users can propose amendments, which will be considered as part of the annual update.

Energy benchmarking: BREEAM

In determining the building's overall performance under this issue, relative weights are assigned to each of the above measures, which reflect the number of BREEAM credits available for each benchmark. This creates a hierarchy. BREEAM has a specific assessment method for data centres. For data centres, in particular those with small office areas attached and low density of occupation, no points are awarded for health and wellbeing of occupants or transport links.

Recommended target achievements are dependent on client's aspirations and should be agreed at an early stage. Table 2 shows the key points to consider when using the BREEAM data centres method.

Table 2:
BREEAM Data Centres 2010 – key points to consider

Management	Life cycle costing for equipment/systems used Register for Considerate Constructors Scheme Commissioning of equipment Building user guide
Health and wellbeing	Lighting levels, zones and controls Thermal comfort and zoning Daylight and views out Indoor air quality Acoustic performance Room depths and glazing
Energy	Power usage effectiveness (PUE) and EPC rating Energy efficient services Sub-metering of energy loads LZC feasibility study BMS
Transport	Proximity to transport facilities Minimise car parking spaces Travel plan
Water	Water leak detection Low flow fittings and rainwater/grey water recycling Water metering
Materials and waste	Re-use of structure/materials Use of materials highly rated in Green Guide Waste storage Responsible sourcing
Land use and ecology	Mitigate construction impacts Improve site ecology and promote biodiversity Appoint ecologist early
Pollution	Minimise NOx emissions from heating sources Provide sustainable urban drainage systems Refrigerant leak detection Minimise watercourse pollution

EPC: energy performance certificate; LZC: low and zero carbon

Energy benchmarking: LEED

The performance rating method prescribes a methodology for establishing a baseline building model. Regulated components include lighting, building envelope systems, heating, ventilating and air conditioning (HVAC) systems, and domestic hot water heating systems. The final determination of performance is based on the percentage improvement of the proposed design model over the baseline design model on the basis of energy cost. LEED does not have a data-centre specific methodology. Table 3 shows the key points to consider when using the LEED method.

Sustainable sites	Minimise pollution from construction activities Provision of transport facilities (cycle racks, showers) Storm water design Parking for shared vehicles Light pollution reduction Heat island effect
Water efficiency	Use water efficient landscaping Use wastewater/rainwater recycling Use low flow water fittings
Energy and atmosphere	Use energy efficient systems Commissioning measurement and verification LZC technologies Purchase green power
Materials and resources	Storage of recyclable waste Re-use of existing structure/materials Construction waste management Use regional materials/responsible sourcing
Indoor environmental quality	Achieve minimum indoor air quality requirements Low emitting materials (carpets, adhesives, paints etc) Maximise daylight and views Water metering Lighting/thermal zones

Table 3:
LEED New Construction 2009 – key points to consider

In a recent case study, the scores achieved using the LEED New Construction 2009 and BREEAM Data Centre 2010 were very similar.

None of the existing benchmarking schemes quantify the environmental impact of the embodied energy of the IT, mechanical and electrical infrastructure of the data centre, which is significant and likely to be higher than the embodied energy associated with the building materials.

6 Engineering systems design

6.1 Electrical services

The data centre is fundamentally dependent on electrical power, not only for the IT equipment it houses, but also the cooling and environmental control of the IT spaces. These electrical systems also have to provide the right level of conditioning, resilience and maintainability by the use of manual and automatic switching from the origin of the supplies down to the IT equipment. Electrical systems therefore comprise a large number of components, each a complex device in its own right, connected to provide a wholesale support system.

In the design of these electrical systems, alternative supply routes and the system's ability to operate under fault or maintenance is of paramount importance where required. A logical process must be employed to analyse the requirement, the loads and resilience (often quoted as tier level), and then adopt a top-down approach – a process of selecting equipment and cabling systems to connect them together. This process is outlined below.

— The utility supply from a distribution network operator (DNO) is the place to start. Depending on the size and location, this can take a few months to a few years to provide and requires negotiation with the DNO. Options have to be considered in terms of the voltage of the supply, the capacity growth, the format, requirement for equipment on site and off site as well as cost and programme. It should be noted that in the UK there are alternative methods of procuring power as well as the local DNO, using an independent DNO (iDNO) or an independent connection provider (ICP) as set out by OFGEM.

— Today, alternative or renewable on-site energy sources are extremely important for meeting reducing targets for energy use and carbon footprint. Full or partial scalable provisions should be considered, together with longevity and payback.

— Working down through the system, key system components have to be selected, their quantities determined by the required redundancy and resilience. The primary ones are listed here.

  • High voltage (HV)/medium voltage (MV) switchgear, its form and switching medium, modularity, fault level etc.

  • Transformers, operating anything from 400 V to 132 kV, two primary types are used, preferably nonflammable oil (midel) and dry type or cast resin.

  • Low voltage (LV) switchgear, its form rating, circuit breaker components, size from 16 A to 6300 A, fault levels and degree of protection.

- A UPS system's primary function is to sustain normal IT loads, and in some high-density sites mechanical plant loads, when utility power is unavailable for whatever reason and allow the on-site generation plant to take on the load. This is achieved by allowing a fall back to a short-term energy source (battery or kinetic store) whilst switching and/or starting up of longer-term forms of energy, such as standby generators. The UPS then also manages the quality of power and provides protection against 'brown outs'. There are two primary methods in providing this function – static electronic circuits and rotating machines with a mechanical drive. When selecting UPS systems, consideration has to be given to the potential configurations, e.g. N, (N+1), (N+2), 2N, 2(N+1) or distributed redundant, their efficiency, their longevity, maintenance requirements, size and operation (see also references to tiering).

- Using UPS systems, isolating transformers and static switches can introduce additional neutral earthing connection points; this needs to be considered when designing the electrical distribution system, with the various modes of: under UPS power (no mains), UPS in bypass with mains and generator. Careful design with the various devices including three- and four-pole disconnectors/protective devices will minimise any future problems of either loss of or parallel of neutral earth paths as well as considering issues relating to maintenance and safe isolation.

- As an alternative power source, the convention is to use an engine and alternator on a common bedplate to provide a standby generator. These can be combined with rotary UPSs to give a diesel rotary UPS or diesel rotary uninterruptible power supply (DRUPS). These all have the appropriate acoustic treatment, fuel system and flues.

- In order to condition the power to the load, power factor correction, active harmonic filters or isolation transformers may be used individually or in combination in the power distribution system.

- The final delivery of power to the IT equipment or racks may be by power distribution units (PDU). These may also include static transfer switches (STS), to be able to switch from different sources, and have detailed electrical system monitoring. Alternatively, power may be delivered via rack power panels (RPP) at the end of rack rows or busbar based systems.

— Utilising the above components, the designer will select the most appropriate types of cable or busbar to provide the required connectivity and power paths. Fundamentally, data centres are either single path (Tier I and II – see Figure 7 and Figure 8) or dual path (Tier III and IV – see Figure 9 and Figure 10 and, as such, components will be connected to ensure the desired level of resilience is achieved.

Figure 7:
Typical Tier I topology

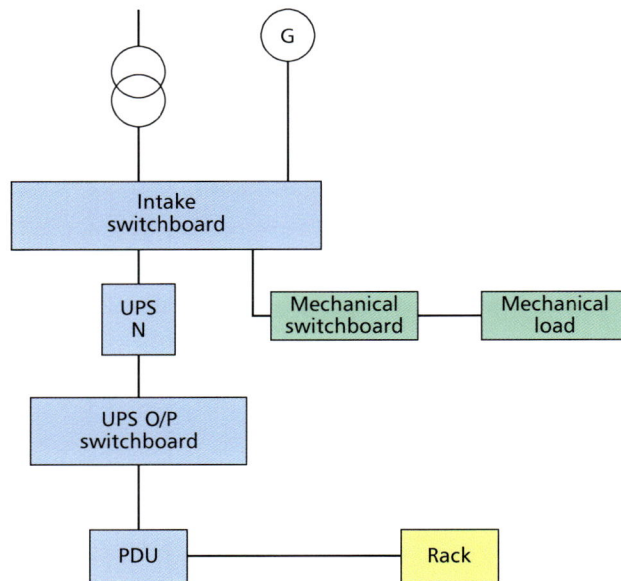

Figure 8:
Typical Tier II topology

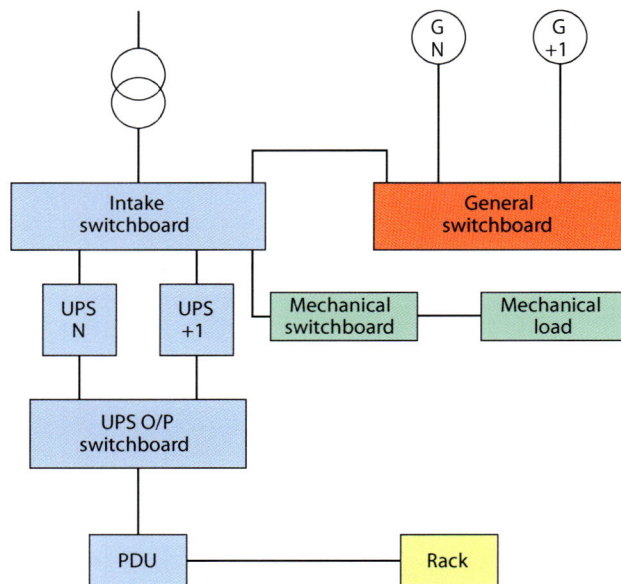

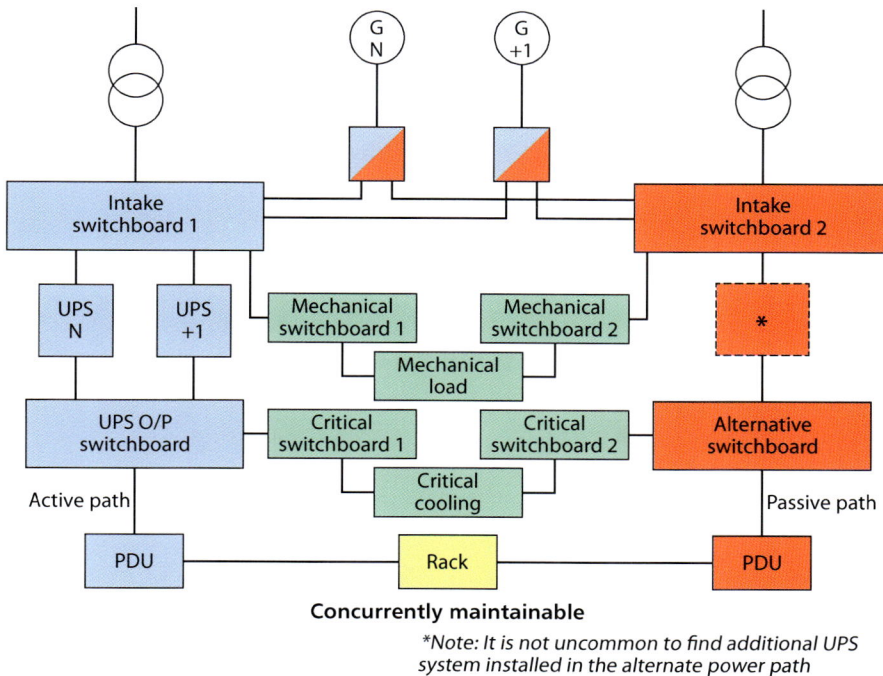

Figure 9:
Typical Tier III topology

Concurrently maintainable

*Note: It is not uncommon to find additional UPS system installed in the alternate power path

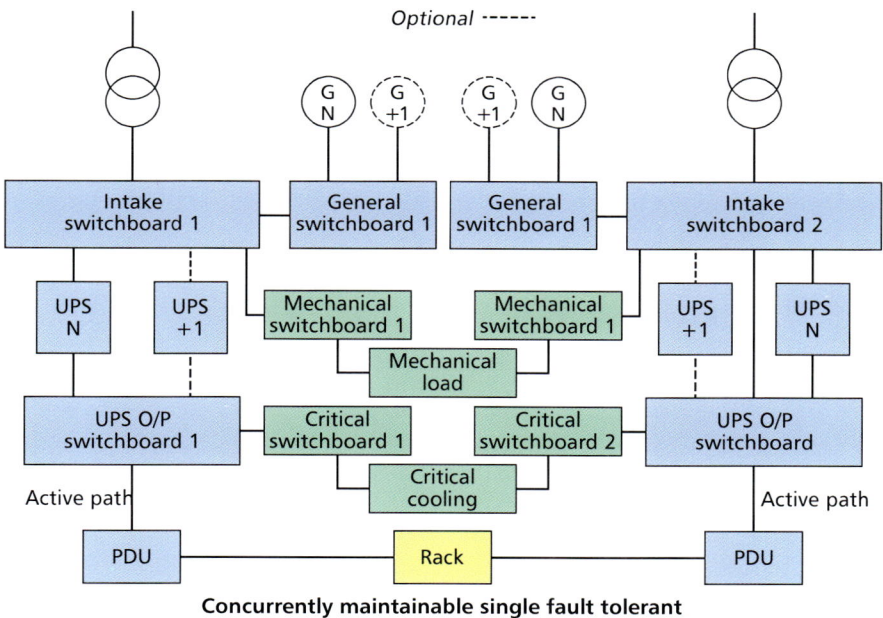

Figure 10:
Typical Tier IV topology

Concurrently maintainable single fault tolerant

— Electrical systems also have to be monitored and sometimes controlled using complex switch relay or programmable logic controllers (PLC) systems. Supervisory control and data acquisition systems (SCADA), energy management system (EMS) and building management system (BMS) may be deployed either individually or as combined systems.

— Other systems are also required, including elements such as earthing, lighting, lightning protection, small power, external lighting, etc.

— Care also has to be taken over off-site and potential on-site sources of electromagnetic interference. Surveys of existing sites and assessment of new designs should be undertaken to define the level of compatibility and what measures may be required to screen data hall spaces.

— All of the components noted will naturally be tested and commissioned in a sequenced and co-ordinated manner, culminating in a combined integrated system test (IST).

LV switchgear and UPS systems should generally be sized so that main busbars do not exceed 5000 A (although 6300 A is possible) and, as such, this limits the size of a single LV UPS system. The hall size and load density dictate the UPS size and exceeding the single system threshold results in multiple UPS systems per hall affecting cost, space and net to gross ratios.

6.2 Mechanical services

There are a range of options to consider when reviewing cooling solutions for the data hall. These need to be assessed in conjunction with the internal criteria, power density and resilience. A number of environmental standards and criteria exist both in Europe and the USA such as the European Telecommunications Standards Institute (ETSI) and the American Society for Heating, Refrigeration and Air-Conditioning Engineers (ASHRAE).

Systems selection criteria:

— temperature

— humidity

— rate of change (both temperature and humidity)

— contamination, air cleanliness and filtration

— plant noise

— resilience

— energy reduction measures

— 'day-one' loads.

Space temperatures in the order of 20–25 °C are considered acceptable with CRAC/CRAH return air temperatures around 10 °C above this (although ASHRAE is proposing 18–27 °C). On this basis, operating chilled water flow and return (F&R) at higher temperatures can be considered, but not necessarily at the expense of refrigeration plant COP. This will achieve less or no latent cooling and possibly less or no need for humidification on internal CRAC/CRAH units.

Humidity ranges of 40–60 per cent are acceptable. Recently, manufacturers of IT equipment have relaxed the range at which their components can operate satisfactorily and this has resulted in increased values of 30–80 per cent relative humidity. As with temperature, the rate of change is important with changes preferably occurring over long periods of time. IT equipment functions best in a stable environment. Temperature swings of less than 3 °C per hour or 5 per cent swing in the humidity level are desirable to ensure optimum performance. To achieve a reduction in energy and water consumption, options other than steam humidification are gaining acceptance. Two current examples are ultrasonic and the use of 'wet mats'. The important point to note is the water used must be sterilised—this can be achieved by using ultraviolet disinfection.

Air cleanliness and the need for good quality filtration is paramount and the external pollution levels have a big impact. Filtration systems should include both primary and secondary filters on air handling systems with European Standards of EU4 and EU6/7 (MERV11/13) class of filters respectively. Room air conditioners (CRAC/CRAH) units would typically be specified as EU4.

The noise generated by the room air conditioning units is not normally a problem as IT equipment may include blade server technology, which produces noise levels around NR70–80 depending on power density. These higher noise levels may result in a requirement to wear ear defenders, increase the output of fire alarm sounders and the installation of flashing fire alarm beacons.

The potential for condensation should not be overlooked, particularly with unoccupied or low-occupancy data halls during cooler periods of the year.

Equipment resilience is a key issue and classification for tiering will determine the degree of standby plant to be provided. A minimum level of acceptable resilience would be (N+1) where N equates to the number of units to meet the load. Where large numbers of CRAC/CRAH units will be installed this is better expressed as a percentage, for example (N+20 per cent) as anywhere up to 50–60 units can be provided depending on their capacity. 'Day-one' loads are generally defined as the period between practical completion and a point in time where design loads are being achieved. IT loads can be very low, especially immediately after handover and normal plant operation will require a degree of modification to achieve economic plant operation while maintaining the desired resilience level. Where systems will be operating at low loads, adequate precautions must be in place to protect variable speed pump motors and chillers to avoid excessive starting or compressors operating at low capacity.

With the increasing demand to achieve energy efficiency and improve PUE, innovative design solutions are now a major consideration in the selection of plant and equipment but these must not compromise resilience. With the acceptance of higher processing temperatures, there has been a noticeable increase in the use of free cooling, all air cooling systems and adiabatic (spray water or wetted mat) cooling. Cooling strategies will be based on end user expectation and designs evolved from this based on various alternatives such as:

— refrigerant options

— space cooling options

— rack load classification

— improving cooling performance

— high-density solutions.

Choice of cooling systems depends on the size of the data centre, energy considerations, control, resilience, capital and operating costs.

The condition of the entering air and its temperature difference across the cabinets is a major factor and a range of alternative options is available to achieve this in an energy efficient manner.

The use of energy-efficient electronically commutated (EC) motors within CRAC/CRAH units is a recent example that, when coupled with floor void pressure control, can provide an energy-efficient method for minimising fan power of the room air conditioning units without compromising control or resilience.

The location of temperature sensors that operate in conjunction with floor void pressure needs to provide an accurate indication. Locations range from measuring CRAC/CRAH leaving air temperature to the face of cabinets in the cold aisle (the latter being the preferred location but secured to the cabinets not freely suspended thermistors, providing adequate provision for safety override is made).

Options on direct expansion (DX) or chilled water are more straightforward but as a rule these will be dictated by the size of the data hall and IT capacity. Space-cooling options are dependent on rack load capacity; various options from perimeter CRAC/CRAH units to in-rack cooling offer a range of possibilities. The cooling option, space configuration, cabinet selection/ arrangement and likely equipment types should all be considered in a holistic manner for design goals to be achieved. The term 'space cooling' is misleading as the critical factor is IT equipment cooling and, since these spaces are largely unoccupied, focusing on space cooling can be an energy waster. The use of hot/cold aisle containment can address the issue of intensified cooling to and heat dissipation from server racks and this is dealt with in more detail in the next section.

Width of the data hall will have a bearing on location and disposition of CRAC/CRAH units. Typically the units have a throw of 10–15 m depending on the depth of the floor and congestion in the raised floor. The positioning of these units is important as they will affect the room size, layout and aspect ratio.

In any new data centre, improving cooling performance is a matter of applying 'good housekeeping' principles to the design process and a clear understanding of air movement within the data hall. Therefore, segregation of cool supply air and warm return air (e.g. by the provision of hot or cold aisle containment and cabinet configuration using baffles) is a must if energy efficient and well-controlled conditions are to be achieved.

Methods for controlling humidity range from steam humidification within a selected number of CRAC/CRAH units to treatment of the fresh air handling plant, which generally serves to pressurise the data hall and prevent the ingress of dust. The M&E services consultant will need to review options that best suit the design.

With the emphasis on conserving resources, the use of 'wetted' systems within fresh air plants has increased in recent years, but given the wide range of tolerance for modern IT equipment the need for humidification should be agreed with the IT team.

If the client is considering a high-density (HD) cabinet solution, e.g. cabinet loads in excess of 25 kW or more, to meet IT needs, it is worth considering a water-cooled cabinet solution. If clients are sceptical about the use of water in a data hall, other methods are available such as DX and carbon dioxide, each having their own advantages and disadvantages.

The need to supply fresh air to the data hall is essential to ensure a positive pressure within the space. Exhaust air can be managed by pressure relief dampers but if fire suppression systems in the form of gas are used then a mechanical extract system with pressure relief dampers will be essential. In the event of gas suppression discharge it will be necessary to vent the space for a period of time to remove the gas and provide safe conditions for entry – this should vent directly to atmosphere.

Mechanical systems resilience

For a number of years the Uptime Institute 'industry standard tier classifications performance' has been the industry standard (see appendix A, which covers Tier I (basics), Tier II (redundant capacity components),

Tier III (concurrently maintainable) and Tier IV (fault tolerant). Tier III and Tier IV requirements for multi-cooling distribution paths and concurrent maintainability of equipment, such as chillers and CRAC/CRAC units, will have a major impact on plant space and service void distribution systems. Figures 11–14 show variations that can be employed on systems designed to Tier III.

Figure 11:
Tier III typical valve arrangement serving an N+1 CRAC system

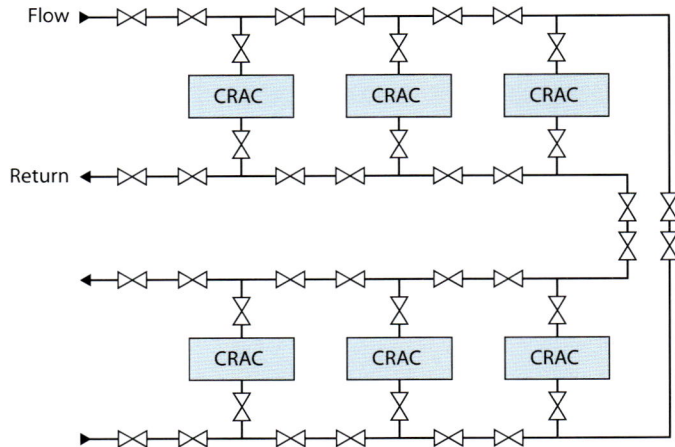

Figure 12:
Tier III typical valve arrangement serving an N+2 CRAC system

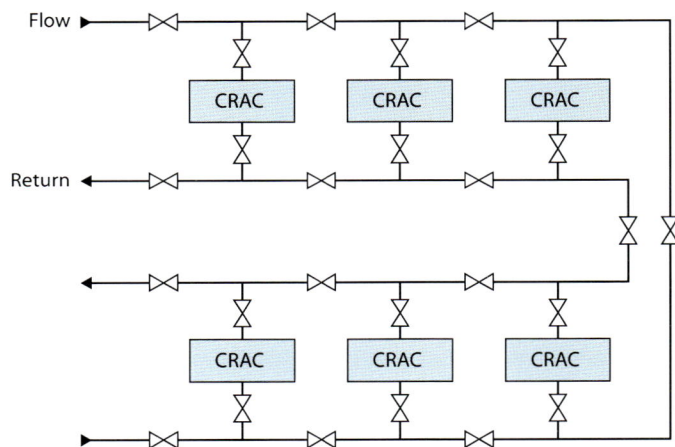

Other systems to control conditions in plant areas, including air conditioning systems, need to be provided. Key areas that should have resilient systems are:

—   UPS equipment rooms

—   UPS battery rooms

—   sub stations and electrical switchrooms

—   security control room.

Normally, mechanical plant rooms, housing pumps, heat exchangers and CRAC/CRAH service corridors, if provided, are mechanically ventilated with no provision for any additional cooling.

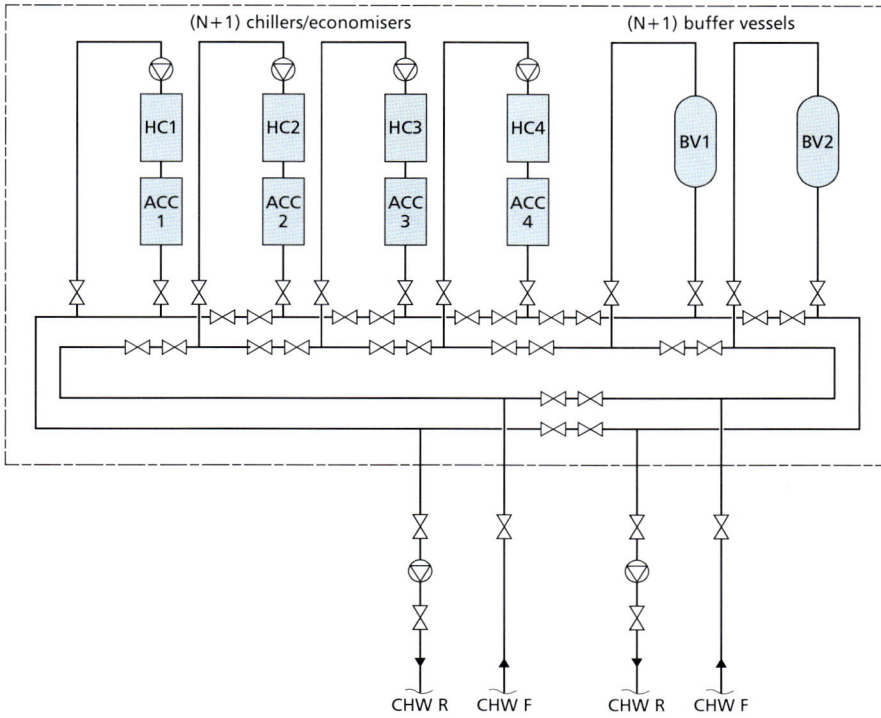

Figure 13:
Tier III typical pipework arrangement serving chillers

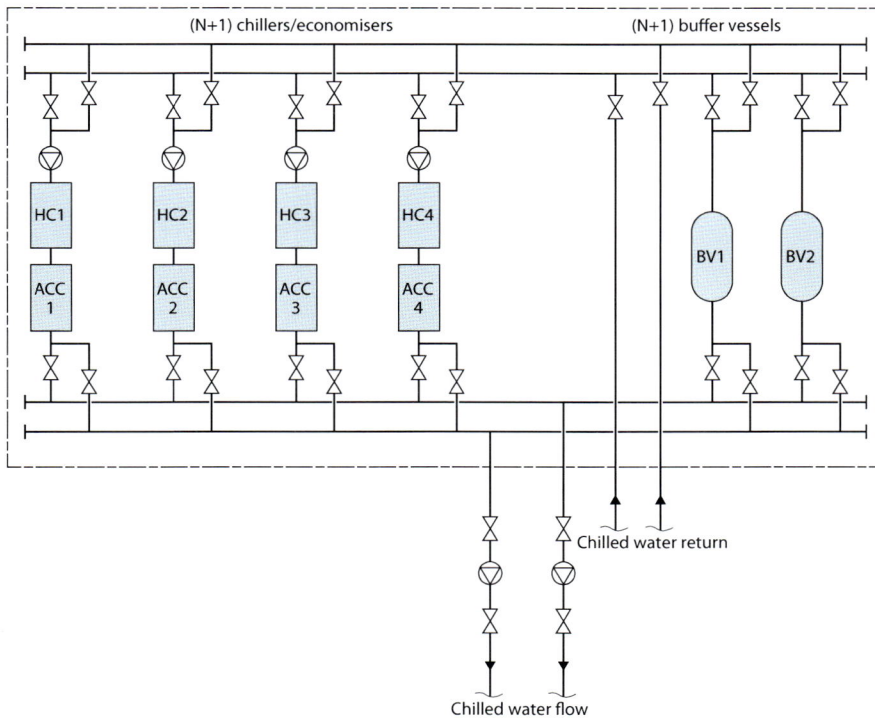

Figure 14:
Tier III alternative pipework arrangement serving chillers

Provision for UPS-backed mechanical services should also be reviewed, particularly in relation to room power density and rate of rise in room temperature.

7   Air management

The purpose of air management is to reduce bypass air to save fan energy and reduce the extent of recirculation so that the set points (air/chilled water) can be increased to achieve higher mechanical efficiencies and allow increased hours of free cooling operation where available. This can be achieved by the following measures.

— Reduce negative flow due to high air velocities near cooling units – this can be achieved by means of perforated baffles under the raised access floor to even out the velocity pressure from the CRAC/CRAH discharge.

— Avoid the use of discharge air scoops, which only serve to increase the velocity of the airflow under the data hall floor. An even disposition of static pressure across the under floor should be achieved, which will assist in achieving a proportional balance through the floor grilles.

— Ensure all floor grilles are installed with opposed blade dampers. This will enable fine tuning of air volume flow though the grilles across the data hall.

— Reduce bypass flow (air cooled that does not reach the IT equipment).

— Reduce recirculation flow (exhaust air from IT equipment that returns directly to the inlet).

Air management is an energy enabler, as on its own it will not save energy. It is critical at both the design stage and during ongoing management throughout the data centre's life in order to be able to utilise the full intended design capacity. One management solution is to use modelling and simulation, such as data-centre infrastructure management (DCIM) or computational fluid dynamics (CFD) tools to plan data centre changes. With poor design or management many data centres encounter overheating and capacity problems well before reaching the intended design capacity.

Figure 15:

CFD illustration of the effects of imbalance between cooling system air supply and IT equipment air demand

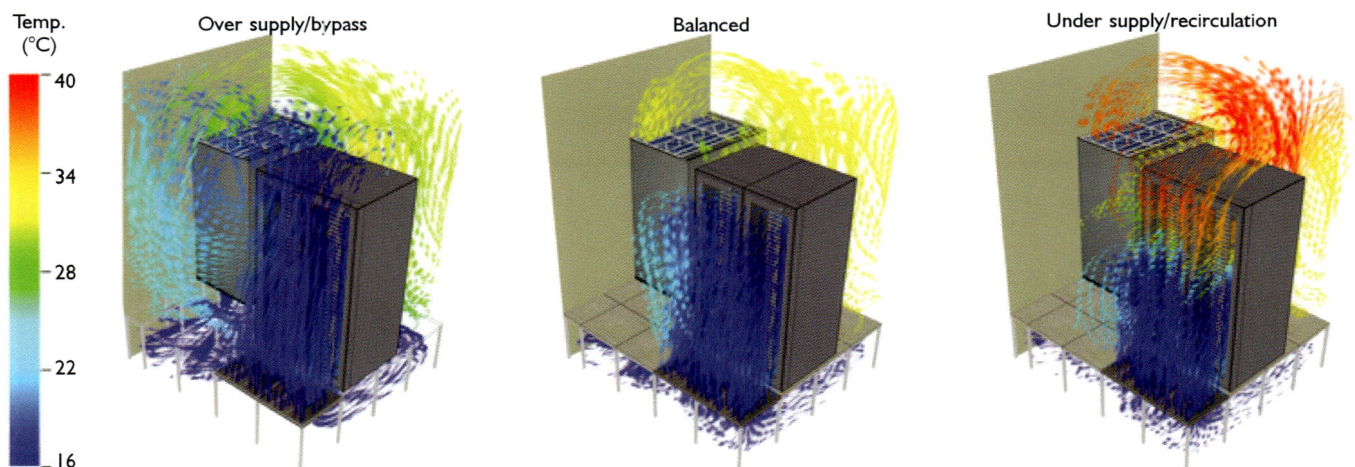

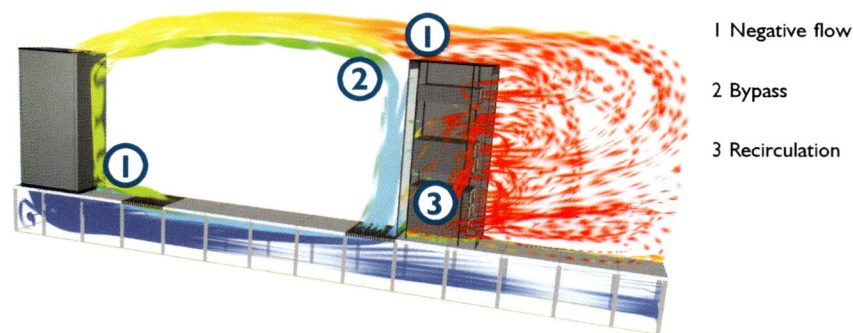

1 Negative flow

2 Bypass

3 Recirculation

Figure 17 illustrates a typical example of legacy data hall airflows shown in sections with the width of the lines indicating the relative volume and colours indicating the temperatures. The extent of bypass and recirculation can be quantified by measuring average inlet and outlet temperatures at the cooling unit and IT equipment and applying mass ratio equations. In the figure, although 1200 kW of cooling is available from the cooling units, only half of this is delivered to the IT equipment as a result of the high level of cold air that is bypassed, resulting in the 800 kW IT load being undersupplied with cold air.

Figure 17:
Legacy data hall airflows

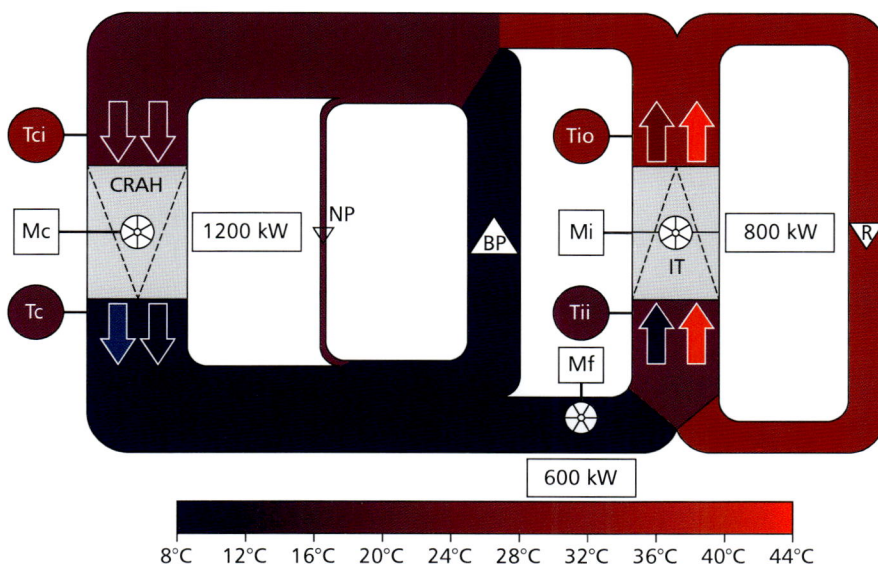

Key:
BP – Bypass
CRAH – Computer room air handler
Mc – CRAH Airflow
Mf – Floor airflow
Mi – IT airflow
NP – Negative pressure

R – Recirculation
Tci – Temperature cooling in
Tco – Temperature cooling out
Tii – Temperature IT equipment in
Tio - Temperature

Physical segregation of hot and cold air streams can minimise bypass and recirculation issues. Methods include cold aisle containment, hot aisle containment and the use of cabinet exhaust chimneys. Figures 18 to 20 show examples.

Figure 18:
Contained hot aisle

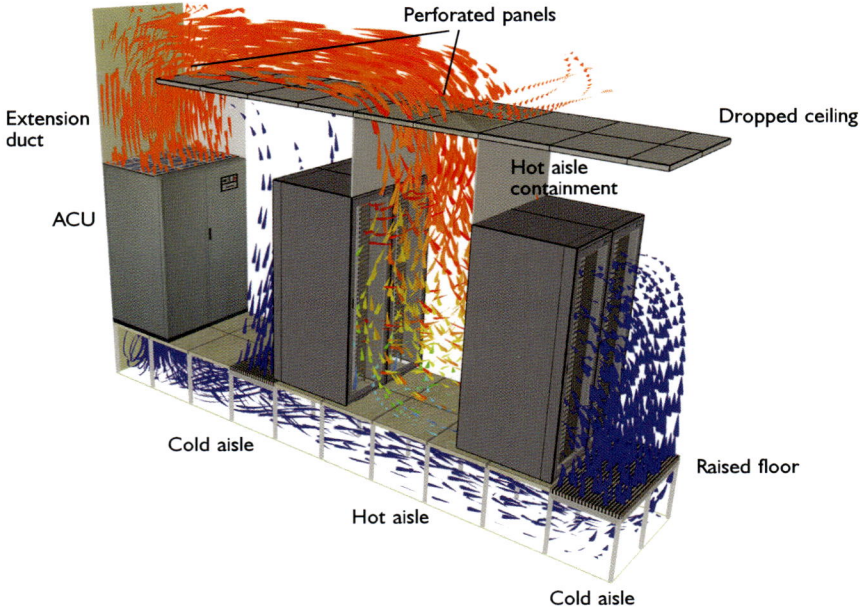

Figure 19:
Contained cold aisle

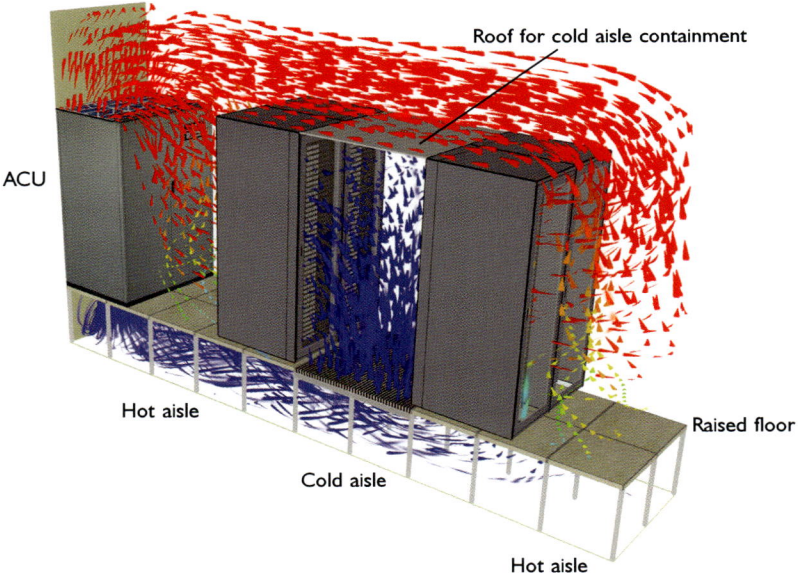

Air conditioning unit (ACU) = CRAC/CRAH units

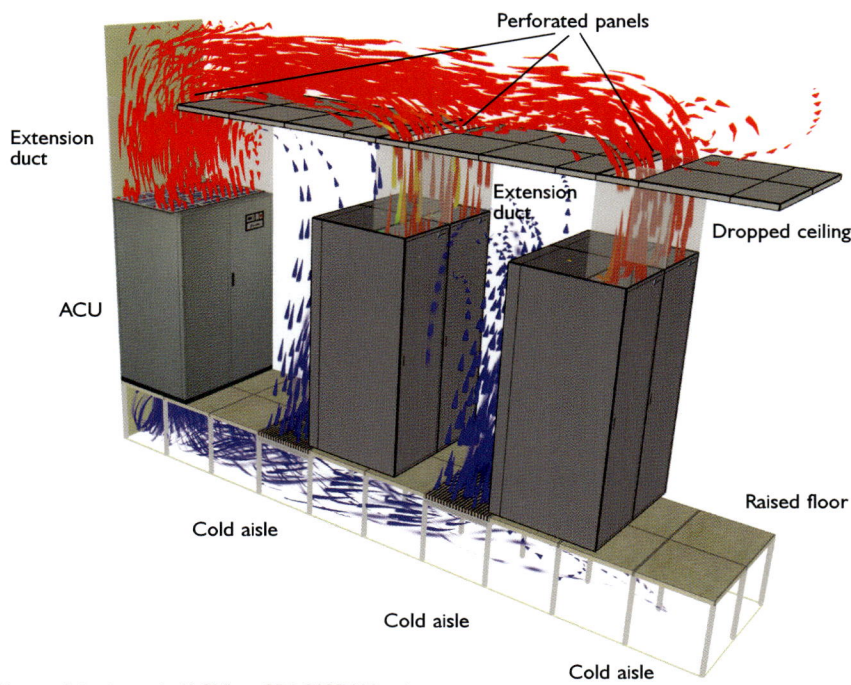

Figure 20:
Chimney exhaust

Air conditioning unit (ACU) = CRAC/CRAH units

The following should be considered when deciding which method is most appropriate:

— co-ordination of fire services including fire detection and suppression systems in relation to isolating a single pod in lieu of an entire data centre hall

— for retrofit case, ease of installation (semi-containment solution with curtains may work best)

— temperature of hot aisle, particularly with future higher IT temperature difference – conditions for operators (this is covered by European legislation).

8 Fire safety

Stakeholder issues

This process requires identifying and engaging with all relevant parties in the decision and approval process and utilising the correct level of technical support for specialist input, including:

— business operations

— statutory authorities

— insurers

— specialists.

Business criticality

What are the risks to business from the threat of fire? A risk assessment will need to review and determine measures required to mitigate risks.

Compliance

This will mean identifying relevant standards and regulations applicable to all aspects of the installation and for the ongoing operation.

Strategy

Develop a fire safety strategy that will, in conjunction with all other operational issues, determine:

— management of fire safety

— evacuation strategy

— fire and smoke control

— fire fighting

— fire protection.

Design and installation

Selecting appropriate solutions will manage risks and therefore meet the requirements of the defined strategy. Consider compliance and capability of systems and resources selected for both installation and continued ongoing lifespan.

Acceptance

Ensure the records and functionality certification of all fire safety measures are correct. Installations will be signed off as appropriate and records of the installation provided. Facility operators are trained and familiar with all services and operating procedures.

Operational

Provide processes and procedures that will be defined to ensure continued statutory compliance of all fire safety measures will be maintained and business risks are properly managed.

8.1   Fire safety systems

Fire strategy

The fire strategy will have a significant impact and must be properly addressed to ensure the people and the business investment are safeguarded and that the installation remains compliant.

A comprehensive strategy should deal with the following issues:
— overview including the impact on the business operation
— management of fire safety arrangements
— evacuation
— fire and smoke control
— firefighting
— fire protection.

Fire protection

Usually, all fire-safety measures and systems are provided in accordance with local relevant standards and building regulations and these will vary on a country-by-country basis. Within the UK BS 6266: 2011 (BSI, 2011) gives recommendations for the protection against fire for electronic equipment installations such as data centres. Risk categories are established based on the type of facility and the level of risk to business continuity.

Fire detection and alarm systems

Fire alarm systems fall broadly into two groups – conventional or analogue addressable systems.

'Conventional' fire alarm systems, in their various forms, have been around for many years and have changed little in terms of technology, although design and reliability have improved significantly. A conventional fire alarm system is often the natural choice for smaller systems or where budget constraints exist.

Analogue addressable systems are typically used in data centres due to the nature of risk and the need for accurate location/detection of alarms. Detectors are wired in a loop around the building with each detector having its own unique 'address'. The system may contain one or more loops depending on its size. The fire control panel 'communicates' with each detector individually and receives a status report, i.e. 'healthy', 'in alarm' or 'in fault', etc.

Sounders may either be conventionally wired or, by using addressable sounders, wired in a 'loop' thereby making considerable savings in terms of cable and labour. A voice alarm system can also be utilised using speakers instead of sounders giving the ability to give a pre-recorded announcement or be used as a public address system. These systems must still be designed and installed to relevant standards for fire safety.

Monitoring and control

Main control panels will be sited in accordance with standards and operate all statutory controls.

Cause and effect

When events occur on any fire system these will need to trigger actions or alerts to any number of fire systems and or building controls. On first knock an immediate investigation in the vicinity of the detector location will take place. These are normally summarised by the production of a cause and effect matrix or diagram (Figure 21 provides an example).

			Fire alarm cause and effect							
			Effect - system operation							
			1st stage sounder activated including remote indication	2nd stage xenon beacon activated	Signal to main fire alarm panel	CRAC units 1&2 shutdown	CRAC units 3&4 shutdown	Secure door release - refer note 3	Signal to gas exting-uishing system	Fresh air supply/ extract ducts close
CER cause	First knock — Single smoke detector activated, air sampling detector activated	VESDA CRAC 1&2			X					X
		VESDA CRAC 3&4			X					X
		Floor void smoke detection	X		X					X
		Room smoke detection	X		X					X
	Second knock — Second smoke detector activated, signal from main fire alarm panel, any manual call point activated, search time expired	VESDA CRAC 1&2			X					
		VESDA CRAC 3&4			X					
		Floor void smoke detection		X	X	X	X		X	X
		Room smoke detection		X	X	X	X		X	X
		Signal from main fire alarm panel	X					X		X
		Manual call point activated (external to area)	X							
		Smoke/duct detector activated (external to area)	X							X
UPS plantroom cause	First knock — First knock - single smoke detector	Floor void smoke detection	X		X					
		Room smoke detection	X		X					
	Second knock — Second smoke detector activated, signal from main fire alarm panel, any manual call point activated, search time expired	Floor void smoke detection		X	X				X	
		Room smoke detection		X	X				X	
		Signal from main fire alarm panel	X					X		
		Manual call point activated (external to area)	X							
		Smoke/duct detector activated (external to area)	X							

VESDA system gives very early fire warning signal to main fire alarm panel to instigate early search period.

CRAC units in CER are shutdown on second stage

All building secure doors, including the CER and UPS plant room doors, are released on second knock, via the main FA Panel.

A confirmed fire elsewhere within the building will release all secure doors, including the CER and UPS plant room doors, via the main FA panel.

Aspirating smoke detection systems (ASD)

Aspirating smoke detectors (ASD) (see Figure 22) and high security sampling detectors (HSSD) are normally provided for the purposes of early warning and are quite different from conventional spot type smoke detectors. They typically comprise a number of small-bore sampling pipes laid out above or below a ceiling in parallel runs, some metres apart. Air or smoke is drawn into the pipework through the holes and onward to a very sensitive smoke detector mounted nearby, using the negative pressure of an aspirator (air pump), with sampling points over the return air path to the CRAC/CRAH units.

Figure 22:

Typical aspirating smoke detection system

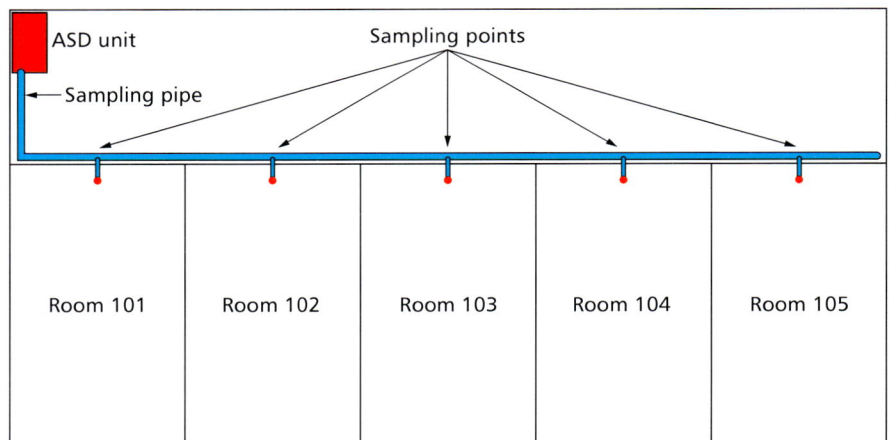

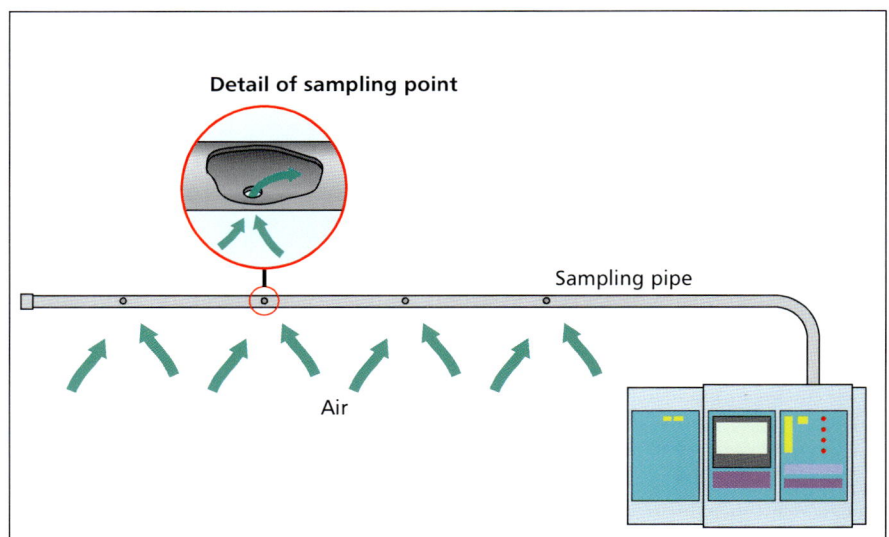

Fire suppression

(*a*) Suppression system types

There are many types of fire suppressing systems that offer varying degrees of protection including:

— sprinklers (wet, dry, pre-action)

— fixed gaseous extinguishing systems

— water mist

— oxygen depletion.

They are nearly always required by the insurance company or at least in part. The type of fire suppression selected may be influenced by the tier rating of the installation, the size and the loss (see appendix A).

(*b*) Types of sprinkler

Wet pipe sprinklers are a common system and are used where there is no risk of freezing.

Dry pipe sprinklers are filled with air under pressure at all times and the water is held back by the control valve. When a sprinkler head opens, the drop in air pressure opens the valve and allows water to flow.

Pre-action sprinklers (like dry pipe systems) have pipes filled with air but water is only let into the pipes when a detector operates (e.g. smoke detectors) (see Figure 23). Pre-action systems are used where it is not acceptable to have the pipes full of water unless there is a fire.

It is not unusual to have a sprinkler installation throughout the general areas and have either a pre-action system installed to data halls/machine rooms or alternative means of fire control/suppression i.e. gas, mist etc or, preferably, both suppression and sprinkler systems.

Figure 23:
Pre-action system detail

(c) Fixed gaseous extinguishing systems

A fixed extinguishing system comprises an automatic and manual means of detecting the early stages of a fire, coupled to a 'fixed' cylinder or cylinders of some form of extinguishant gas, as opposed to hand operated 'mobile' extinguishers commonly found in most buildings.

The detection and control equipment comprise a control panel, to which the automatic smoke detectors or sensors report and alarm outputs to warn occupants of a fire or fault condition or an impending discharge of gas. Outputs affect the controlled shutdown of electrical distribution boards, air conditioning or fresh air input plant and the closing of any smoke control dampers. See Figure 24 for a typical fire suppression system.

In the event of a gas discharge of chemical gases, there can be large thermal changes within the space. The main gaseous types, suitable for occupied rooms are therefore inert gases, such as Inergen, Argonite, argon, or a chemically manufactured gases, FM200, 3M™ or Novec 1230™.

Both inert and manufactured gases require the protected space to be evacuated before discharge.

Figure 24:
Typical fire suppression system

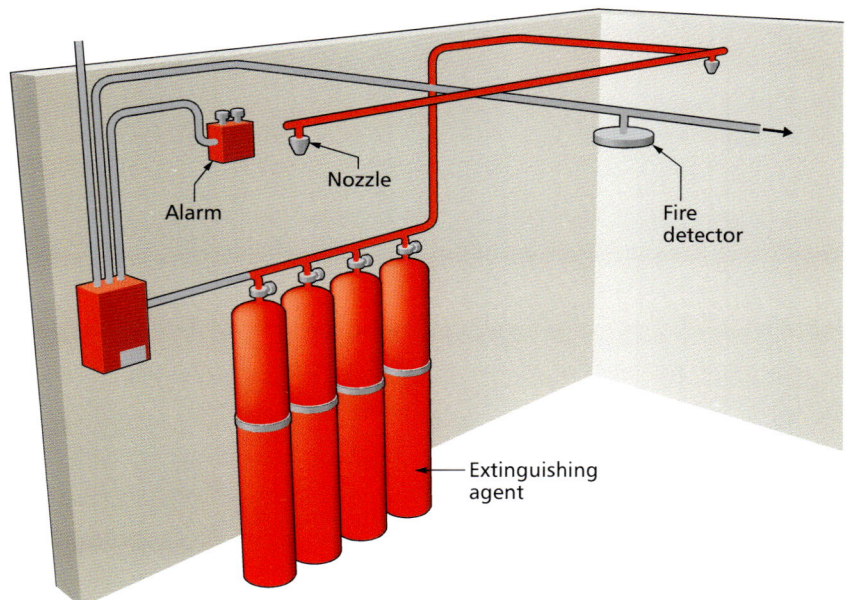

(*d*) Water mist systems

Water mist systems typically require less water to suppress a fire than traditional sprinkler systems. The water mist method extinguishes a fire by rapidly absorbing its heat and excluding oxygen from the fire at source – all achieved by discharging very fine droplets of water. The mist can penetrate hard-to-reach areas as the fine water droplets are 'airborne' and are drawn to the fire via the natural induction of air.

Water mists are also appropriate for fire suppression of diesel engines whether in containers or in combined rooms.

Water mist's fire suppression ability has been recognised for many years. This performance is due to the large total surface area of the droplets combined with the high speed at which they convert to steam and absorb the energy of the fire.

It should be noted that any form of hot or cold aisle containment can have a big influence on the fire system's performance and this should be considered when designing installations.

(e) Permanent fire prevention with oxygen reduction

Fire prevention through oxygen reduction (sometimes referred to as OxyReduct) is a method that has been used in a number of applications since the 1990s. However, in recent years, this fire protection technology has experienced a rapid development. Compared with the traditional fire extinguishing and fire suppression methods, the oxygen reduction technology provides an active approach to the fire prevention goal, creating an atmosphere where a fire simply can not develop from a source of ignition.

9 Network provision

Adequate, appropriate and expandable network provision – both wired and wireless (although wireless is less common as bandwidth makes copper and fibre far more cost effective) is a prerequisite for any data centre and must be planned at the earliest opportunity. Provision must be made for three sets of networking needs and these will need to have growth potential in terms of numbers of circuits, data volumes and data transmission speeds:

— connections into the data centre from buildings on site, off site on WAN/LAN and from the internet

— connection of those devices, e.g. servers, disc arrays, backup devices, housed within the data centre itself often in increasingly densely populated racks

— connections into the data centre from the many client devices within the building housing the data centre.

See Figure 25 for an example of a schematic network structure.

A key challenge of designing a solution for today is building in sufficient flexibility to cater for emerging technologies and future growth in demand. In terms of technology, Institute of Electrical and Electronics Engineers (IEEE) standards task groups are evolving standards, e.g. for 40 and 100 Gb, but these are a long way from being ratified. As a result the designer is often reliant on a few leading manufacturers who are bringing products to market that are largely second guessing the standards.

Figure 25:
Schematic network structures typical of most installations

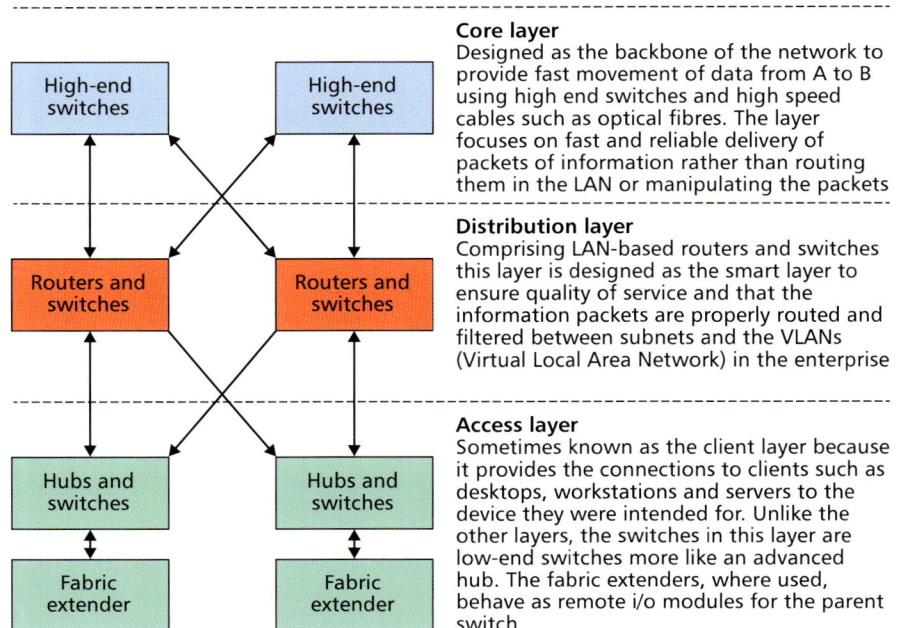

Core layer
Designed as the backbone of the network to provide fast movement of data from A to B using high end switches and high speed cables such as optical fibres. The layer focuses on fast and reliable delivery of packets of information rather than routing them in the LAN or manipulating the packets

Distribution layer
Comprising LAN-based routers and switches this layer is designed as the smart layer to ensure quality of service and that the information packets are properly routed and filtered between subnets and the VLANs (Virtual Local Area Network) in the enterprise

Access layer
Sometimes known as the client layer because it provides the connections to clients such as desktops, workstations and servers to the device they were intended for. Unlike the other layers, the switches in this layer are low-end switches more like an advanced hub. The fabric extenders, where used, behave as remote i/o modules for the parent switch

Designs for UK-based data centres including communications, network hardware and cabling should consider design standards such as BS 6701: *Telecommunications Equipment and Telecommunications Cabling*, TIA: 942 *Telecommunications Infrastructure Standard for Data Centers* and BS EN 50173-5: *Information Technology. Generic Cabling Systems. Data Centres.* BS EN 50173-5 applies throughout Europe. Although other standards do exist, notably in North America and Australia, compliance in Europe can be more difficult.

For data centres, the IT and network infrastructure should consider the following points, taking account of its tier rating:

— the overall strategy for future phasing and proofing should be defined and agreed with the customers to ensure the proper capacity is reserved

— availability of internet service providers (ISP), carriers and points of presence (POP) to the locality

— entry routes, diversity, demarcation points, segregation, security requirements etc

— degree of resilience and routing within the data centre

— the overall and detailed physical limits and restrictions on cabling, in terms of bend radii, length restrictions, capacity in containment, packing and capacity planning for the future

— layout of the networking racks and cabinets in relation to the cabinets served; co-ordinate with the cooling provision, power supplies and lighting systems

— size (H, W, D) of racks and co-ordination with flooring systems and access

— backbone distribution design between incoming points, telecommunication rooms, campus connections, other data halls and network concentration points

— arrangement and delivery of structured cabling to cabinets and other IT equipment in a logical format that allows for later flexibility

— specification to TIA 942 (TIA, 2012) of fibre and copper based cabling using standards such as OM1 to 3 and OS1 and category 5, 5e, 6 or 7 respectively; for copper cables, the use of unshielded and shielded twisted pair cabling should be reviewed

— presentation of ports, whether fibre standard connectors (SC) or lucent connectors (LC) or copper (RJ45) at servers and equipment and networking equipment

— use of pre-terminated cabling, either single or double ended

— the overall management of the cabling systems including final delivery, hierarchy and access should be planned and co-ordinated with other services and the building fabric

— the vertical alignment possibilities for backbone accessibility and circuit distribution when the data centre is located in a multi-storey building

— types of containment required, e.g. ladder rack, basket, conduits, specialist fibre containment etc, ensuring that bend radii, capacity and connections properly match the initial and future requirements; decisions on the delivery of containment and cabling above or below computer racks should be taken in conjunction with other M&E services and, in particular, the impact on cooling should be considered

— separation of the cable routes from other services to standards such as BS EN 50174 (BSI, 2009)

— design of network closets and telecommunication rooms in terms of carrier capacity and access to other parts of the site or campus around the data centre

— how the data centre and its network should be controlled and managed typically using an on-site network operation centre (NOC) or bridge.

The results of these considerations should provide an overall strategic IT network plan facilitating flexibility and growth over the lifetime of the data centre. It must cater for expected growth in demand for transactional interactive access, the increased use of virtualisation, the increased use of multi-media using high definition, expected and unexpected peak network workloads and high availability 24/7. All these will require both faster bandwidth and more bandwidth, which changing IT and network technology will satisfy over time.

10   Security

10.1   General security

In assessing the physical security arrangements, key considerations are:

— perimeter

— internal areas

— technical space

— cabinet access

— data security and network protection

— encryption.

Threat and risk assessment

The quantity, location and type of security equipment installed will need to be commensurate to the level of residual risks that require further reduction or mitigation. Such risks would be identified as part of a threat and risk assessment.

The following notes are based on a typical site assuming medium level of risk.

Operational security, physical security, electronic security

A secure site will result from the right balance of operational, physical and electronic security systems. Other contributing factors are location, people, processes, resilience of power, communications and the availability of the data stored at the data centre.

— Operational security consists of day to day procedures and emergency planning and, as such, will not be considered within this section.

— Physical security takes the form of barriers to deter and delay forcible entry into the site and building, such as gates, wall construction, doors and windows.

— Electronic security is based on systems that enable forcible attacks to be detected at the earliest opportunity and dealt with appropriately (they can also prevent access to authorised users).

Figure 26:

Typical site perimeter layout

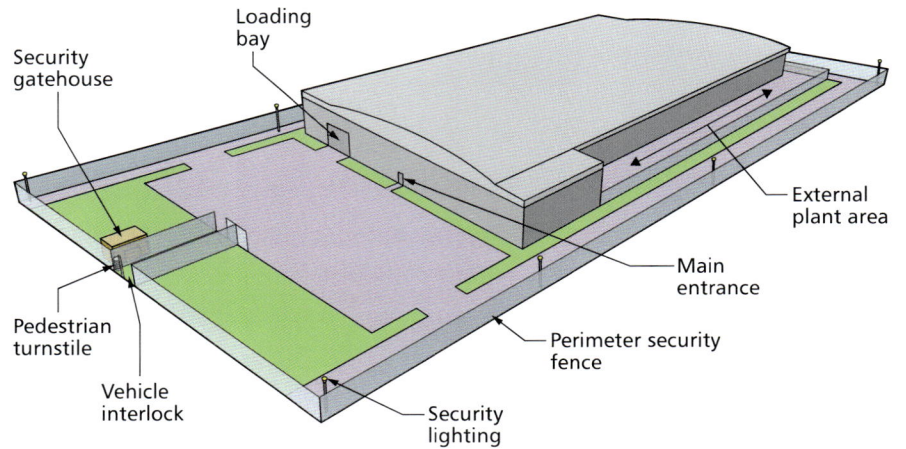

Figure labels: Loading bay, Security gatehouse, External plant area, Main entrance, Perimeter security fence, Security lighting, Vehicle interlock, Pedestrian turnstile

10.2 Monitoring and control

Areas to consider in relation to monitoring and control are outlined here (see Figure 26 for an example site).

Parking

Where possible, parking should be outside of the secure site perimeter to eliminate vehicle risks and the security logistics associated with vehicle screening.

Fencing

The complete site should be enclosed by a fencing system. Typically it will be 3 m in height and may require the addition of a topping such as barbed or electrified wire.

Electronic security

A perimeter intruder detection system (PIDS) should be installed at the site perimeter fence to detect attempted intrusion, climbing or cutting.

Security gatehouse

For larger sites with frequent vehicle and pedestrian traffic a security gatehouse should be deployed to enable designated security staff to efficiently process entry and exit.

Vehicle and pedestrian access gates

Automated vehicle gates that match the perimeter fence should be provided to control access. The practicality of swing or sliding gates and the requirements for crash-rated equipment should be considered. They should form a vehicle entrance interlock of approximately 15 m in length between an outer and inner gate.

Electronic security

Entry into the outer and inner gates should be restricted and logged by combination of card and PIN readers and video intercom units. Pedestrian gates should be provided for access and should be automated and arranged to eliminate tailgating with audio visual communication.

Video surveillance cameras

Video surveillance cameras will be required around the site perimeter to allow for proactive monitoring. The majority of cameras will be mounted on cabinet base columns, which may require planning consent. Each cabinet base of a camera column will contain control equipment wired back to a security equipment room via fibre optic or structured cabling. If a high lux level of white lighting is not available infrared lighting will need to be installed. Areas to consider in relation to internal monitoring and control are outlined here (see Figure 27 for an example internal security layout).

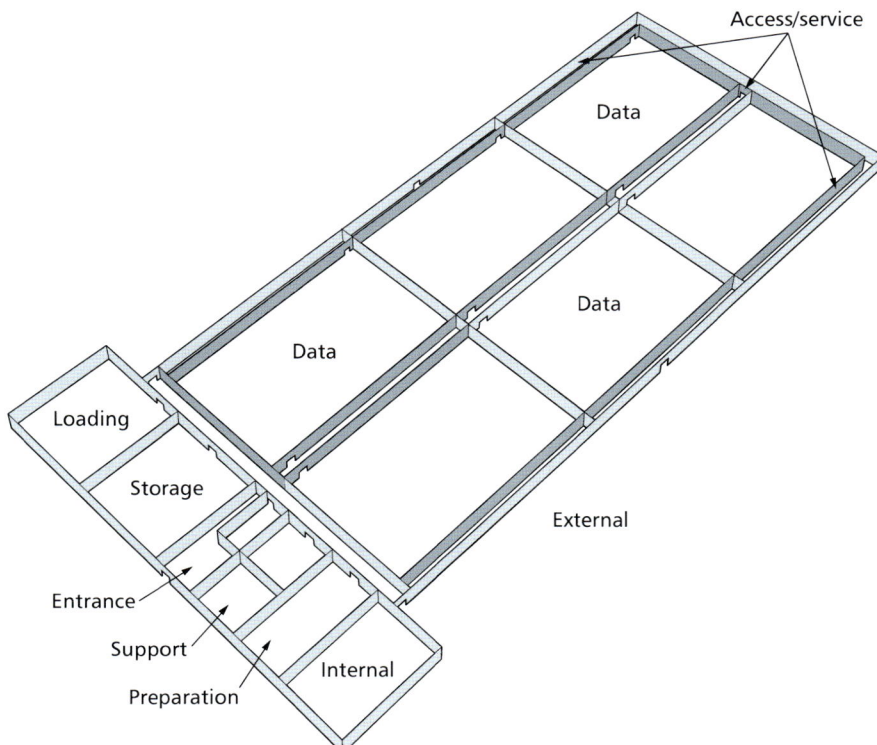

Figure 27:
Typical internal security layout

Main entrance lobby

— Is the facility is to be manned 24/7? If so, other considerations and decisions will have to be made.

— The walls and doors of the internal wall between the main entrance lobby and the main centre should be constructed from manual attack rated materials and span from slab to slab.

— The external main entrance door should be controlled by electronic locking using combination card and PIN readers and a video intercom unit.

— The main entrance waiting area should be constructed to 'meet and greet' at a PayPoint-style window of the security control room before allowing access directly to the main corridor.

— Surveillance cameras must be installed to provide general surveillance of the main entrance waiting area.

Main thoroughfare corridor

— Entry into the corridor from the main entrance lobby should be controlled by a door that eliminates tailgating such as a circle lock tube. Air locks should be considered.

— Surveillance cameras should be located in positions where they can produce images of all people entering the corridor.

Service corridors, service rooms and data halls

— The walls and doors should be constructed from manual attack rated materials and span from slab to slab.

— Depending on the level of security, any aperture greater than 250 mm^2 should be fitted with mesh or grilles.

— Entry should be restricted by electronic locking and combination card and PIN reader with egress by card reader. Entry into data halls should be restricted by a door that eliminates tailgating. Entry into the unit should be restricted by a combination card and PIN and fingerprint reader with egress from the data hall by card reader (reference should be made to the appropriate intrusion level standards dictated by the brief).

— A double-leaf equipment door should be provided, opening from the inside only and restricted by electronic locking.

- All doors should be fitted with magnetic door contacts.

- Surveillance cameras should be located to view every door and provide images in which people can be identified.

Loading bay

- The loading bay should be constructed as a holding area for deliveries.

- All walls and doors of the loading bay should provide a high degree of resistance to manual attack.

- Typically, the loading bay would be fitted with an external roller shutter resistant to manual attack.

- All external door(s) of the loading bay should only be opening from inside the building.

- A video intercom unit should be installed outside the loading bay to summon assistance and aid secure communications.

- Internal doors of the loading bay should be restricted by electronic locking and card readers. Air locks should be considered.

- The loading bay external door should have an interlock arrangement with all internal doors off the loading bay so that they cannot be open at the same time.

- Surveillance cameras should be located within the loading bay and provide recognition of individuals with other areas covered by general surveillance.

11   Building construction

11.1   Professional team

Normally, key numbers engaged on a data centre forming the team would consist of:

— client (stakeholder)

— project manager

— cost manager/quantity surveyor

— architect/space planner/interior designer

— mechanical and electrical services consultants

— fire engineering consultants

— commissioning management

— structural engineers

— security consultants

— planning supervisor, construction design and management (CDM).

The team would be supported by additional specialists such as:

— legal advisors

— planning consultants

— utilities infrastructure

— acoustic

— landscape architects

— environmental consultant

— local government state or county authorities

— ecology/archaeology consultant

— energy and sustainability consultant.

11.2 Project procurement

Consider how the project will be procured:

— form and type of contract

— lump sum tender

— two stage tendering

— design and build

— partnering or framework agreement.

11.3 Construction programme

The construction programme will vary with the size, type and classification of the facility. A construction programme should be drawn up to demonstrate clearly that sufficient time has been afforded to pre-construction activities, procurement, and the construction, commissioning and close-out. Close-out is considered the most important element of any data centre building. The programme should be regularly updated in terms of progress at weekly and monthly intervals, which is beneficial for both review and progress reporting. See Figure 28 for an example programme.

11.4 Construction costs and cost plan

From the business case a number of cost drivers will be developed including such items as mechanical electrical and plumbing (MEP) system solutions, modularisation of build and adjacencies. The cost plan will then be developed in relation to the evolving brief and design development.

Typically, cost plans would be produced at various stages. To be effective the cost plan should capture all the potential project expenditure including professional fees, a risk allowance and local or government taxes. Cost will also be affected by the programme if the project overruns. This will ensure there are no surprises and the project is being delivered within budget.

Figure 28:
Simplified programme for a medium size (2000 m²) data centre

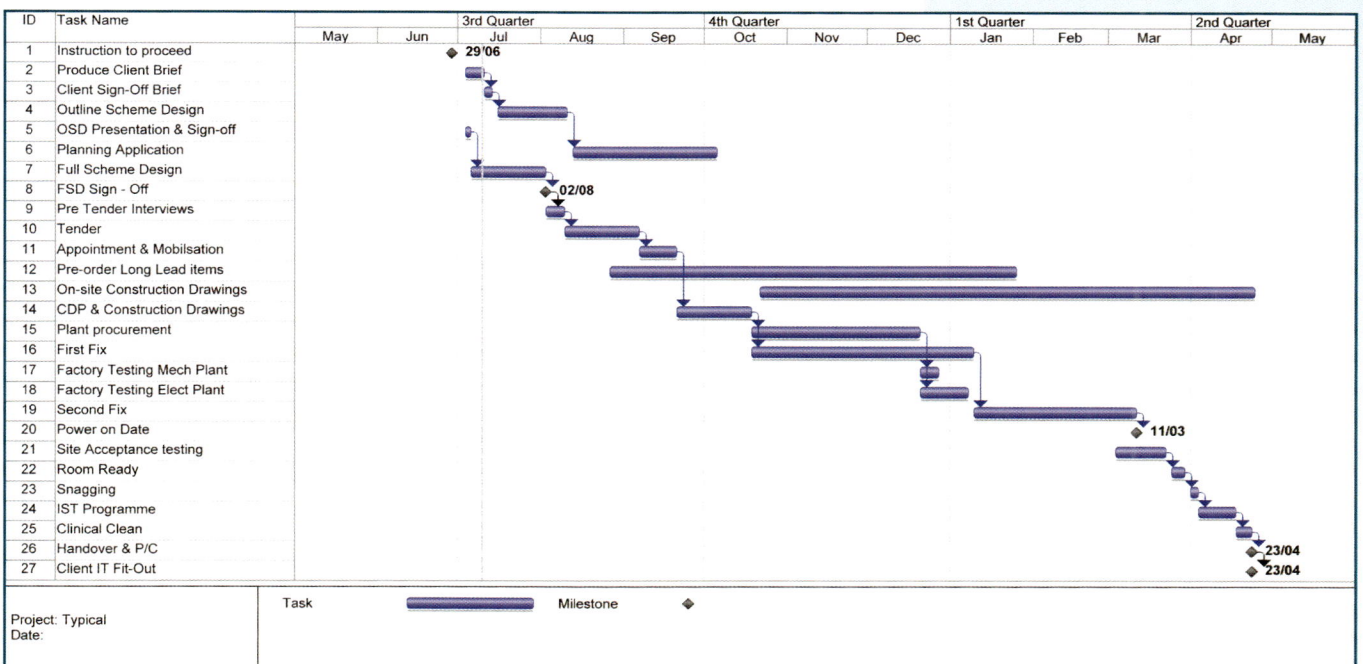

ID	Task Name
1	Instruction to proceed
2	Produce Client Brief
3	Client Sign-Off Brief
4	Outline Scheme Design
5	OSD Presentation & Sign-off
6	Planning Application
7	Full Scheme Design
8	FSD Sign - Off
9	Pre Tender Interviews
10	Tender
11	Appointment & Mobilsation
12	Pre-order Long Lead items
13	On-site Construction Drawings
14	CDP & Construction Drawings
15	Plant procurement
16	First Fix
17	Factory Testing Mech Plant
18	Factory Testing Elect Plant
19	Second Fix
20	Power on Date
21	Site Acceptance testing
22	Room Ready
23	Snagging
24	IST Programme
25	Clinical Clean
26	Handover & P/C
27	Client IT Fit-Out

Project: Typical
Date:

Task   Milestone

Figure 29:

Indicative trend between tier rating and cost (fit-out of existing building) based on average UK costs for 2010

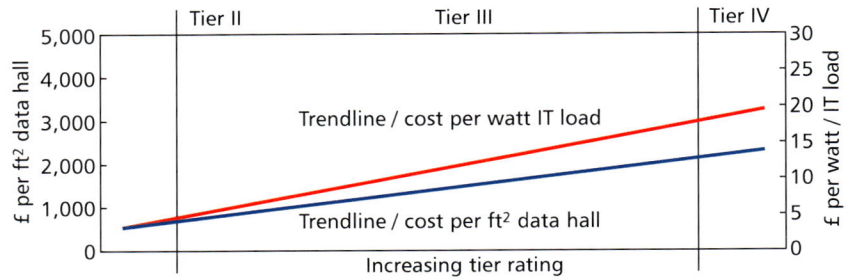

Figure 30:

Indicative trend between tier rating and cost (new build data centres) based on average UK costs for 2010

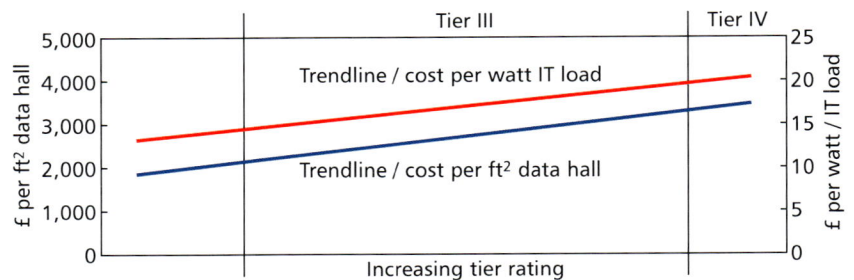

Figures 29 and 30, based on the UK construction industry, provide cost per ft² and IT cost per watt for both new and existing buildings and the relationship between the Uptime Institute tier rating and escalating costs.

They are based on cost per ft² and the tier rating. The tier rating is based on definitions set out by the Uptime Institute, which defines the site infrastructure performance on the basis of reliability and resilience. See appendix A for a clearer understanding of tier ratings.

The cost breakdown for a typical project within the UK excluding IT equipment is around:

— 24% construction

— 54% electrical services

— 22% mechanical services.

11.5   Mobilisation

Under the Construction (Design and Management) Regulations 2007 in the UK, clients are required to provide the principal contractor with a mobilisation period; other regulations and conditions will apply in other countries. This commences when the principal contractor plans and prepares activities prior to commencing works on site and is dependent on communication, co-operation and good management principles.

1.6   Procurement schedule

This is important when preparing the construction programme, with key factors being:

— impact on deliveries of long lead items such as generator, UPS, transformers and fire suppression systems

— major plant selection based on client global partner agreements

— specialist suppliers for uniquely specified equipment.

The procurement schedule will define the procurement requirements, with provision for holiday shutdown periods for the project, and how it will be managed in conjunction with the professional team from developing procurement documentation through to contract completion.

11.7   Health and safety plan

In the UK, the principal contractor has to produce a health and safety plan, but this may vary from country to country. This outlines the key arrangements ensuring the works are carried out safely. Construction should not commence until an adequate plan is in place and has been approved by the client. The health and safety plan should comprise two parts:

— Part A: the development section

— Part B: the site action plan.

Both parts should be continuously developed throughout the project.

11.8   Site facilities and procedures

Site planning should ensure safe and efficient operation with site accommodation located adjacent to the site entrance to control vehicle and pedestrian access (see further reading and weblinks).

There should be adequate provision on site for:

— welfare facilities (toilet, washing and meal breaks)

— site storage (for materials and equipment)

— personal protection equipment (PPE), e.g. hard hats, boots, 'high-vis' vests, safety goggles and gloves etc.

Site facilities must comply with any local codes or regulations.

11.9 Site procedures

Operatives are normally required to be inducted in site procedures and rules before carrying out any works. This induction process will also include members of the professional team, clients and visitors.

Part of the construction process in the UK requires risk assessments to ensure safe working for operatives. These are structured reviews of work activities, which identify hazards. In response to risk assessments, method statements that describe construction and maintenance activities will be issued to ensure they are carried out safely.

11.10 Construction activities

— Programme monitoring is an effective management tool by which the principal contractor records progress for each activity against the construction programme.

— Regular site meetings will take place between client/professional team/ sub-contractors, to review all technical, operational and commercial matters that relate to the project.

— Design workshops are a collaborative approach to deal with elements of the design where input is required from the construction team. As construction progresses this may generate 'change requests' which will require sign-off. Contractor requests for information (RFIs) will be raised where clarification is required.

— Commissioning should be considered at the beginning of the project. The consultant's/client's brief should be reviewed to ensure that the commissioning plan satisfies the specification and the design intent. Adequate provision should be made in the construction programme to allow for comprehensive integrated systems testing (IST). This may also identify clashes within the construction programme.

— Technical submissions are a safety mechanism for all parties. These are documents prepared by the contractor and issued to the professional team to confirm technical details of the project's plant and equipment prior to ordering.

— A factory acceptance test (FAT) of plant and equipment is critical for key items of the plant to validate performance data and identify and eliminate any problems at source. A FAT specification, which describes the tests to be adopted, is normally issued and witnessed by the consultant/client. These tests will often be at 'full' and 'part' load.

— Beneficial access to areas of the site and phased handover are often a client requirement and will need to be programmed in.

— Site security is essential if vandalism and theft etc, which may cause major disruption to the programme, are to be avoided. It is important that detailed security arrangements are in place.

— Fire safety will be of paramount importance when working within a 'live' data-centre environment and this will also require the strict control of 'hot works' and permits to work.

— The monitoring and control of both dust and vibration will also be a major consideration in any 'live' environments.

11.11   Energy and materials management

Data centres are large energy consumers and considerable effort is being placed on minimising this during and after construction.

Companies and data centre facilities providers have to be active in demonstrating their commitment to energy and sustainability. As a result, a number of schemes to judge their performance are available.

In the UK, these environmental certification schemes developed by the Building Research Establishment (BRE) encompass:

— Environmental Profiles Certification Scheme

— Responsible Sourcing of Construction Products

— Microgeneration Certification Scheme (MCS).

In part, these are also covered by:

— BREEAM

— LEED

— Green Star.

11.12   Close-out

Strictly speaking this should be referred to as 'completion management', which is covered in more detail in section 12. This process covers a number of key stages prior to commencing IT wiring and server population.

Testing and witnessing

This will be managed by the completion management team and includes factory and site testing plus any ISTs, including 'load (heat bank) tests' required by the client and other established procedures.

Documentation

Test sheets will be produced as evidence that the testing processes have been properly performed witnessed and signed off prior to any scenario testing commonly referred to as integrated system testing (IST). Operation and maintenance (O&M) manuals will, as a minimum, include plant, systems and equipment, plant overview and operating instructions, a plant replacement strategy and 'as built' record drawings.

Snagging procedures

Recognised arrangements need to be in place to enable any snags identified by the client's team to be cleared and signed off on time. Provision should also be made in the overall construction programme for snagging remediation.

Clinical clean

Standards are available for room-clean procedures, e.g. BS EN ISO 14644-1 (BSI, 1999–2006); it is important for the client IT team to agree levels of cleanliness and have these signed off prior to commencing any IT wiring.

Training

Training is integral to and part of the O&M instruction programme. Validating the content of this would normally be undertaken by members of the professional team, such as the services consultants. The training, along with any potential 'soft landing' arrangements, would be provided to either a client's in-house team or FM service provider to ensure a good understanding and safe operation of all systems. This would include but not be limited to:

— BMS interfaces and critical leak detection systems, relative humidity and temperature alarms

— fire alarm and suppression systems

— CRAC/CRAH operating regimes.

Soft landings

Soft landing procedures ensure a smooth transition from construction to operation of the building systems and services. A good explanation of this is the UK Building Services Research and Information Association document BG4/2009: *Soft Landing Framework* (BSRIA, 2009), which provides clear guidance from inception to aftercare, ensuring a smooth transition.

12   Completion management

Competition management is probably the most important part of the construction programme. At this point, the mechanical and electrical systems installed are virtually complete and the programme for commissioning testing and setting to work commences.

The role of the completion management team is critical in ensuring the plant and system are properly tested and this includes the integrated systems testing. Figure 31 is an example of a typical matrix that defines roles and responsibilities.

The four main topic headings for correct execution of the completion management phase of the project are listed here.

Commissioning management plan

The key to the success of the commissioning management phase is to ensure a completion management plan is produced that clearly identifies all key deliverables.

Key deliverables are those that have to be fully proven, tested and signed off prior to any ISTs for mission critical systems. These key deliverables can best be summarised as follows:

— electrical installation complete and tested

— generators tested using load banks for the IT load

— building management system (BMS)

— fire systems (full cause and effect testing)

— UPS

— chilled water system (fully loaded and performance tested)

— O&M manuals

— maintenance preparedness.

Figure 31:
Completion management matrix

No.	Name	Client Rep/ consultant	Main contractor fit out	Package contractor	Commission-ing manager	Validation manager
	Terms of reference matrix	**IST and commissioning roles and responsibilities**				
1	Provide a resource chart for the duration of the project	All	All	All	All	All
2	Perform commissionability review of the design	Review	Review	Review	Lead	Review
3	Production of commissioning and pre-commissioning method statements	Review	Review	Lead	Lead	Review
4	Off-site testing (FATs)	Review	Review	Lead	Witness	Witness
5	Project-specific test sheets Performa	Witness	Review	Lead	Witness	Witness
6	Produce a mechanical and electrical load testing logistics plan for levels 3, 4 and 5 (IST)	Review	Review	Review	Lead	Review
7	Produce and issue commissioning plan (including commissioning network/logic)	Review	Review	Review	Lead	Review
8	Produce the commissioning schedules	Review	Review	Review	Lead	Review
9	Develop a contract commissioning programme	Review	Lead	Assist	Lead	Review
10	Set up and maintain a commissioning issues/risk register	Review	Review	Review	Lead	Review
11	Produce an asset register (plant schedule)	Review	Review	Lead	Lead	Review
12	Issue a site testing/witnessing/documentation protocol	Review	Review	Review	Lead	Review
13	Develop a training plan for the facility maintenance team	Assist	Review	Review	Lead	Review
14	General on site testing (SATs)	Witness	Witness	Produce	Witness	Witness
15	Critical systems site testing (SATs) and general testing	Witness	Witness	Produce	Witness	Witness
16	Maintain a comprehensive testing and commissioning file	Review	Review	Review	Lead	Review
17	Undertake liaison with building control (re FA cause and effect)	Assist	Lead	Review	Assist	Review
18	Prepare a building control witnessing file in accordance with the requirements for the issue of an occupation certificate	Assist	Assist	Review	Lead	Review
19	Develop a performance monitoring strategy (trend logging) during the level 5 (performance and interface testing)	Lead	Review	Review	Lead	Review
20	Manage weekly commissioning meetings and take minutes	Witness	Witness	Witness	Lead	Witness
21	Issue weekly commissioning reports with progress graph	Review	Lead	Assist	Lead	Produce
22	Produce and issue a two-week commissioning activity look ahead	Assist	Lead	Assist	Lead	Witness
23	Actively assist the principal contractor in the production of recovery programmes (if required)	Assist	Lead	Assist	Lead	Review
24	Issue a final practical completion report based on the key deliverables and commissioning deliverables	Review	Review	Review	Lead	Review
25	Sign-off all issues arising out of the commissioning process (commissioning issues)	Review	Review	Review	Lead	Witness
26	Manage and control the O&M manual submissions and comment on the content	Review	Review	Lead	Review	Review
27	Manage the as built drawing issue	Review	Review	Lead	Review	Review
28	Manage the preparation of the energy log book	Review	Review	Lead	Witness	Review
29	Produce the ASHRAE level 4 performance testing strategy	Assist	Review	Assist	Lead	Assist
30	Produce the high level IST test schedule	Lead	Review	Review	Lead	Review
31	Produce the IST test documentation and test plan	Assist	Review	Assist	Lead	Assist
32	Manage the IST testing process	Witness	Witness	Witness	Lead	Assist
33	Training (proof of competency)	Review	Assist	Review	Lead	Review
34	Life safety system	Witness	Witness	Produce	Witness	Witness
35	Witness critical systems levels 2, 3 and 4 testing	Witness	Witness	Produce	Witness	Witness
36	Witness non critical systems levels 2, 3 and 4 testing	Witness	Witness	Produce	Witness	Witness
37	Flushing	Review	Witness	Produce	Witness	Witness

Key to actions:

- **Review item** – comment and assist if necessary
- **Assist with item**
- **Witness** – item to be witnessed/audited by action owner. Item to be signed off if necessary
- **Lead** – action owner to lead/co-ordinate/manage item; acceptance test certificates to be issued if necessary
- **Produce** – action owner to produce relevant documentation for item, which may include implementing tests or reports

Commissioning management brief

The role of the commissioning manager will be of a predominantly managerial nature, rather than merely technical fault finding, administrative or a witnessing body. Their primary function is to actively manage and control the commissioning works that meet the programme, ensure commissioning works are undertaken in a controlled manner and prove performance to design failure scenarios (IST).

Having established the key project deliverables, it is essential that the commissioning management brief is properly specified to ensure all elements of the project commissioning and testing are covered.

The most critical aspects for the success of the commissioning and testing are:

— co-ordination and provision of method statements to ensure that at all times the full extent of trades' work is known, along with their interaction with others

— co-ordination of FAT to ensure that all delivered equipment is fully compliant

— co-ordination of benchmarking to ensure consistent high quality of installation

— preparation of programme integrating all aspects of the services

— establishment of issue workshops to brainstorm potential pinch points.

Testing

Testing and commissioning are the most important parts of this process. Commissioning managers will be employed and will manage the commissioning process to oversee and manage the inspection, testing and specialist commissioning process. The tests should demonstrate what constitutes previously defined pass or failure. Typically, this is:

— *Level 1*: FAT

— *Level 2*: supplier/sub contractor installation tests

— *Level 3*: full witness and demonstration testing of installation/equipment to client/consultants

— *Level 4*: testing of interfaces between different systems, i.e. UPS/generators/BMS etc, to demonstrate functionality of systems and prove design

— *Level 5*: IST.

The principal contractor may be asked to carry out a planned failure of the mains power supply to the building, often referred to a 'black building test', to prove resilience in the event of power failure. This is often allied with IST.

Integrated systems testing is the final demonstration to verify the facility is resilient under all possible failure scenarios. The IST plan is normally produced by the commissioning manager and is developed in accordance with the project brief. During this process the commissioning manager is responsible for overall control of the IST and sign-off.

Discussions to agree the definition and programme dates of 'room ready' are other important aspects of this procedure.

A 'clinical clean' of the technical environment is essential prior to handover to the client's IT team. Specifications for this may vary, e.g. BS EN ISO 14644-1 (BSI, 1999).

Levels 1 to 5 have been based on the ASHRAE document *Design Considerations for Datacom Equipment Centers* (ASHRAE, 2009).

Integrated systems testing

The objective of IST is to ensure that the critical systems supporting the operation and function of the data hall operate as designed and provide power and cooling without interruption following the loss of mains power and/or failure of redundant support systems.

IST therefore brings together the culmination of all the individual systems commissioning into a complete independent test showing each system's ability to both support and sustain the operations of the data hall during potential fault/failure scenarios.

It is therefore important that sufficient time is allocated in the programme planning and execution of this vital element of the project.

A successful systems integration test should consider potential extreme conditions and combinations and is necessary to establish the long-term resilient operation of the data hall. This is also useful for staff training.

Close-out documentation

To ensure that the facility can be correctly maintained and operated it is imperative that the project build and commissioning is fully documented to a high standard.

The following documents should be completed, reviewed and approved prior to completion of the IST (certain exceptions to this can be allowed as long as it has been agreed by the professional team, albeit relaxation of these requirements can be seen as a failure of the management of the entire process):

— asset register

— Conformité Européenne (CE) certification certificates

— building log book (or addendum if part of existing facility)

— O&M manuals plus record drawings

— individual systems commissioning test packs

— client training log and supporting data

— soft landings procedures.

13 Operation and management

13.1 Facilities management

How will the facility be managed after practical completion? Main points to consider are:

— delivery method: will it be internally operated or outsourced?

— tiered or self delivery

— recruitment and mobilisation.

Having invested a substantial amount of money in the critical environment to house companies' critical assets it is essential that the infrastructure that supports these assets is maintained and operated to a high standard to prolong the life of the infrastructure and prevent the risk of failure. Clearly written processes and detailed procedures will be essential for operating and maintaining the facility (see Figure 32 for a cause and effect escalation diagram). These will ensure that a collaborative approach is followed by everyone that works in and around the space. The following procedures should be in place as a minimum to operate the facility.

— Physical access and security: this should be owned and controlled by the data centre manager. Regular audits should be undertaken to ensure that only key members of staff have access to the facility. Non-resident engineers and visitors requiring access to the facility should issue a detailed site specific risk assessment and method statement well in advance of their required visit date to allow the relevant permits to be raised to cover the works to be undertaken.

— Environmental management and control: this will include temperature, humidity and cleanliness of the space and must be continuously monitored to ensure that the facility is operating within its design capacity.

— Change management: any installation or removal of an item of equipment has the potential to interrupt the operation of the business and all parties must understand this before the work takes place. This includes all maintenance activities in and around the critical infrastructure. Detailed method statements must be produced and approved before any work is undertaken.

— Capacity management: approval of any new installation of equipment must be approved by the maintenance engineer who is monitoring the power and cooling consumption within the space. This will ensure that there are no overload situations introduced into the data centre over the lifetime of the facility or equipment 'hot spots'.

— Health and safety: a clear health and safety policy must be in place to ensure the wellbeing of everyone working in and around the data centre. Detailed site-specific risk assessments must be undertaken to ensure that everyone who needs to work within the space can do safely.

— Standard operating procedures: these should be produced for every item of equipment that supports the facility. They should be documented, held in a central place and made available to the engineering maintenance team.

— Emergency operating procedure: emergency operating procedures should be produced to cover any failure scenario that may occur within the facility. These procedures should be practised on a regular basis as a walk-through exercise to familiarise the engineering operations team with the operation of the plant in an emergency situation.

— Cable management: it is essential that a strict cable management procedure is in place to ensure redundant cables are removed once they are no longer required. This ensures a clear air path is maintained under the raised access floor to allow conditioned air to pass through the cabinets and ensure that sufficient cable ways are available for new equipment to be installed throughout the lifetime of the facility.

Figure 32:
Cause and effect escalation diagram

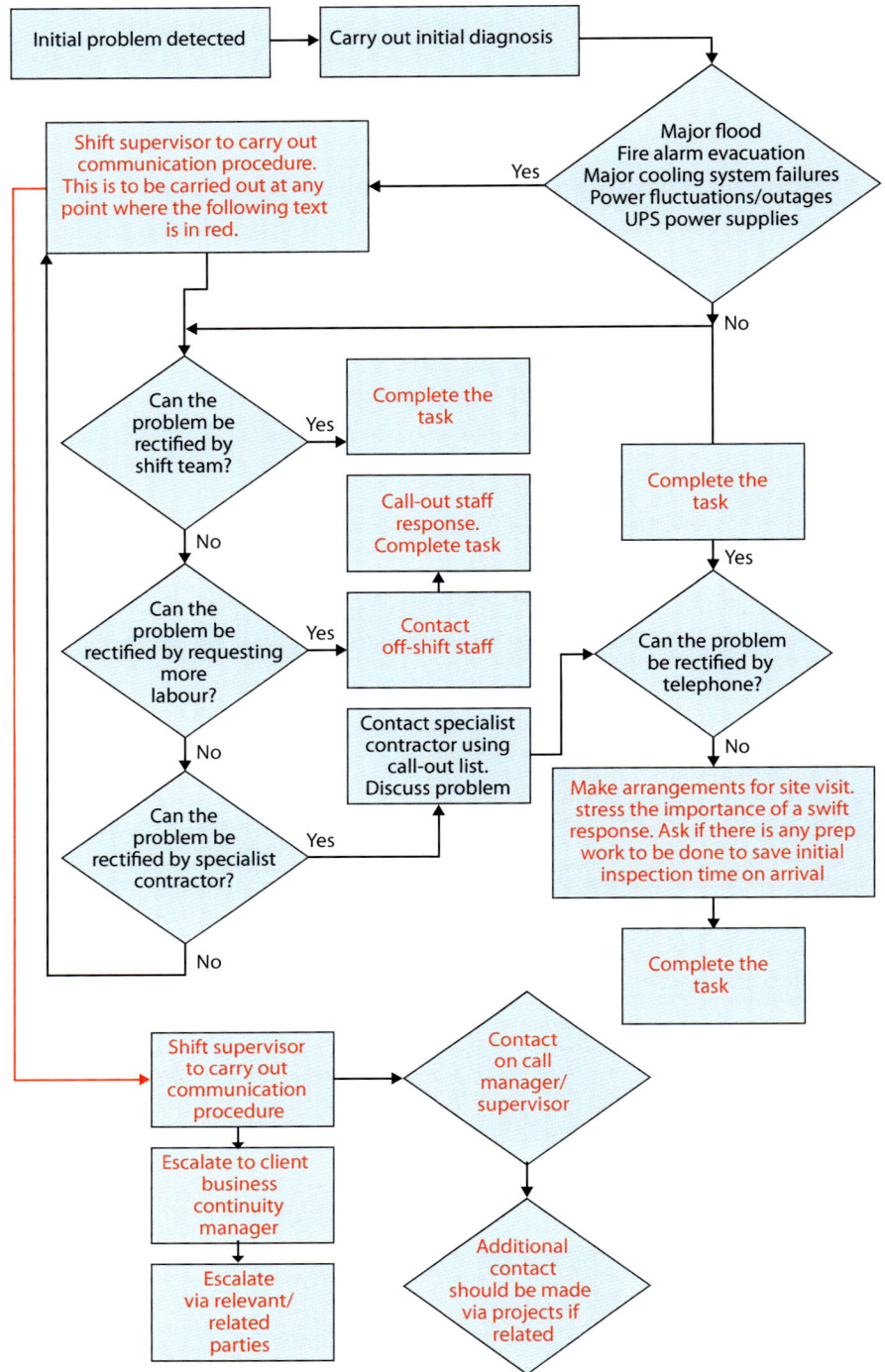

| Initial problem detected | → | Carry out initial diagnosis |

Major flood
Fire alarm evacuation
Major cooling system failures
Power fluctuations/outages
UPS power supplies

Yes →

Shift supervisor to carry out communication procedure. This is to be carried out at any point where the following text is in red.

No

Complete the task

Can the problem be rectified by shift team?
Yes → Complete the task
No

Can the problem be rectified by requesting more labour?
Yes → Contact off-shift staff → Call-out staff response. Complete task
No

Can the problem be rectified by specialist contractor?
Yes → Contact specialist contractor using call-out list. Discuss problem
No

Can the problem be rectified by telephone?
Yes
No → Make arrangements for site visit. stress the importance of a swift response. Ask if there is any prep work to be done to save initial inspection time on arrival

Complete the task

Shift supervisor to carry out communication procedure → Contact on call manager/supervisor

Escalate to client business continuity manager

Escalate via relevant/related parties

Additional contact should be made via projects if related

14 Ongoing developments

14.1 General

With the continual rise of energy costs the data-centre market is set for significant review and change. The recent advent of virtualisation and the introduction of cloud computing has resulted in significantly higher IT equipment utilisation in providing greater efficiencies and this needs to be matched by the systems that support them. The items discussed in this section refer to technologies that (at the time of writing this document) are being considered more frequently in new designs, upgrades and operations.

14.2 Facility size and capacity

Some estimates for the cost in energy of running a facility are set to outstrip the capital cost of construction by a factor of five over the life of the facility. Consequently, traditional views are being questioned.

Is size an advantage or disadvantage? Whilst large data centres provide the comfort of space for growth, operational flexibility and apparent economies of scale (such as fewer larger cooling systems) it is probably appropriate to consider options of building smaller data centres or building modular data centres. More providers are now offering modular solutions, from containerised solutions to larger prefabricated designs. An alternative is to still build a large facility, but construct it with smaller halls in a modular way so that the infrastructure can be added in smaller steps and utilised more efficiently.

14.3 Increased tolerance of higher ambient temperatures

The acceptance of higher air temperatures (ASHRAE recommended range is 18–27 °C, extended allowed range 15–32.2 °C) reaching the equipment is a result of changing technology and a recognition that with good air management and the adoption of hot/cold aisle containment the low supply air temperatures of the past are no longer necessary. However, care must be taken if existing IT legacy equipment is installed as this may not be suitable for operating at these higher conditions. Adopting higher supply air temperature can enable significant reductions in energy consumption and carbon emissions and pave the way for PUEs approaching 1. Designs aiming for a PUE of 1.1–1.3 are becoming more frequent. It should be noted, however, that adopting higher temperatures could significantly reduce the period of successful operation following a cooling system or utility power failure, so when adopting these higher temperatures the designer should consider the failure scenario due to reduced thermal 'head room'.

14.4 Reduced air volume requirement by IT equipment

Designs used to allow for air volumes of 70 litre/s per kW of IT load however the two factors listed below mean less air and higher temperature difference is possible in design, which facilitates higher return temperatures and the potential for energy saving.

— IT manufacturers have reduced air requirements for cooling the equipment by adopting more effective internal cooling air management designs and alternative techniques such as cold plates and heat pipes.

— Intelligent IT fan control means that fan speeds can be reduced at low utilisation and at lower inlet temperatures.

14.5 Increased rack heat densities

Higher equipment densities in racks using all air-cooled solutions are being contemplated and, in a few instances, installed. For this to be effective, the installations will almost certainly have to use physical containment between hot and cold air streams as illustrated in section 7:

— cold aisle containment

— hot aisle containment

— rack chimney containment

— rack row top duct containment.

Such high densities can also be cooled by providing liquid-based cooling direct to the rack or ultimately to the individual chassis or even the chip.

In the case of air-based systems, these higher heat densities tend to result in higher exhaust temperatures at the equipment. If this hot air is not allowed to mix with cool room air before reaching the cooling coils/return air system, the cooling system can run at higher temperatures and increased efficiency is likely and greater free cooling is possible. The issue of mixing undesirable air is often addressed using containment.

Further actions to reduce carbon dioxide emissions per unit of electrical energy produced or better still carbon dioxide per unit of compute in addition to lower PUE will be:

— the use of IT equipment that allows energy savings in other ways, e.g. video conferencing, therefore establishing IT efficiency metrics will be very useful to the industry

— higher IT equipment densities and utilisation will inevitably increase the temperature difference across servers or result in higher cooling flow rates. These higher temperature differences will also make partial free cooling more viable due to the larger number of hours when outdoor conditions will be below that of the IT equipment.

As efficiencies of data centres improve significantly over the next years, the next focus is likely to be on IT and M&E embodied energy (more impact than building materials)—this is the energy required to extract the raw materials, process and transport products to site. Whilst data centres are much more energy intensive than standard buildings, they are also intense with high embodied energy materials, e.g. silicon, aluminium, copper, steel.

14.6 Alternative cooling distribution media

Although air is an easy way of distributing cooling around the data centre, its low density and relatively low specific heat capacity mean that it must be moved in large quantities requiring significant fan energy, not to mention space, for the large distribution cooling and equipment associated with air delivery. Alternative approaches such as the use of carbon dioxide, refrigerant and liquid cooling delivered directly to the cabinet and potentially the chassis or even chip offer the potential for efficient cooling and even higher equipment power densities.

14.7 Cooling infrastructure

The pressure on energy efficiency and reducing carbon emissions means there is considerable pressure to increase the efficiency of the external cooling plant. The intention is clear—to maximise the extent of using 'free cooling'. It is important to recognise that the success of free-cooling approaches will be significantly affected by choice of internal cooling strategy, alongside the selection and management of IT equipment. In particular, increased tolerance by IT equipment of higher air temperatures combined with better airflow management substantially increases the opportunity for energy saving and, indeed, free cooling. However, it is vital to recognise that energy saving (when compared with conventional designs) can only be achieved if the external cooling systems are modified either by adopting free cooling or higher chilled water/operating temperatures.

The simplest approach would be to use outdoor air directly, assuming you can get sufficient air volume into the building to meet the IT load, but the following challenges must be addressed.

— Is the climate sufficient all year round (temperature and humidity)? If not, a mechanical cooling/conditioning system may be required for use when it is not sufficient.

— What are the filtration requirements to adequately clean the outdoor air?

— Is there sufficient access for large air volumes to be supplied from outside to the data hall and for the hot air to be exhausted from the data hall to outside?

14.8   Indirect free cooling using air-to-air heat exchangers

If external air cannot be used directly, one solution is to use an air to air heat exchanger using outdoor air to cool the re-circulated air from the data centre. Again the viability of this approach will in many cases be limited by the local climate and the space for air access and the heat exchangers. There is increasing attention on using the evaporation process (adiabatic) to add additional cooling that could also be applied to indirect air-side free cooling. This could take the form of plate heat exchangers, thermal wheels or run-around coils.

14.9   Adiabatic/evaporative cooling

This is not a new technique but it is becoming more attractive as a result of the focus on saving energy. Evaporative cooling uses the latent heat of evaporation to achieve lower air or water temperatures. When moisture is sprayed into air that is not saturated, some of the moisture will be evaporated to make moist air; this uses heat and, hence, cools the air. The down side is that if sprayed directly into the air being used to cool the data centre, the air in the data centre may become excessively humid. There are two ways of avoiding this:

— an air-to-air heat exchanger spraying into the non-data-centre side of the heat exchanger, getting lower air temperatures on both sides

— spraying water onto a cooling coil reduces the temperature of water or other liquid-based cooling, including condensing refrigerants.

14.10   Cooling recovery devices

When considering the use of all outdoor air-cooling systems, cooling recovery by means of the following methods may be a worthwhile exercise if using CRAC units on recirculation:

— thermal wheels

— matrix plate heat exchangers (sometimes referred to as recuperators)

— run-around coils.

14.11 Renewable energy

The growth in energy demand has resulted in a number of innovative options being adopted to offset this. Broadly, they fall into three or four categories defined as renewable sources of energy:

— solar

— wind

— tidal/wave

— geothermal

— biomass (although it could be argued biomass is a byproduct resulting from solar activity).

Renewables can make a contribution to saving energy and carbon however the quantity of renewable is often constrained by the geography or site constraints. Some examples are listed here.

— Relatively small footprint of a data centre compared to its power consumption. The data centre having average power densities of say 1.0 kW/m^2 and photo voltaics (PV) for example delivering an average of 0.01 kW/m^2.

— Proximity to residential property might preclude the use of wind power due to noise and flicker issues.

— Limited geothermal output of say 400 kW for a borehole with a data centre load of tens of megawatts.

Cogeneration and combined cooling heat and power (CCHP) can deliver a carbon saving, but the effectiveness of the solution is dependent on site specific opportunities and a balance between space on site, capital, costs, maintenance costs and load profile. Introducing these systems may also preclude the use of other technologies like fresh air cooling that would potentially save even more carbon than cogeneration/CCHP. The full spectrum of energy-saving/renewable technologies should be examined at concept design stage to determine the most effective strategy for saving carbon and energy and ultimately running costs for the data centre.

The other major consideration is the resilience of a data centre and its dependence on a reliable power source to continuously support the facility under all conditions.

Although these options appear attractive, they offer little opportunity to exploit their full potential for use within data-centre design other than a modest contribution from PV sources for possibly supplementing external lighting. This, coupled with a fall in the price for the cells and government incentives in promoting these, does offer some consolation.

14.12　Cold air as a renewable source of cooling

Renewable energy is normally a term associated with tidal, wind, solar or geothermal technologies most of which are limited by site constraints or are intermittent in nature. In temperate climates 'cool' fresh air is available for most of the year, is renewable and, with the right application, can save a considerable amount of energy and carbon. Technologies now exist to utilise fresh air for cooling (without compressor driven refrigeration) for the majority of the year 'topped up' with adiabatic/water-mist cooling for extremes of temperature. These technologies can deliver PUEs of less than 1.25. In warmer climates fresh air can still be utilised to provide a considerable proportion of cooling sometimes 'topped up' with a combination of adiabatic/water-mist and compressor driven cooling.

14.13　Fuel cells

A fuel cell is an electrochemical device that converts the chemical energy of a reaction into electrical energy, with heat produced as a byproduct. The application would only be considered if the intention was to generate power on site using gaseous fuel as the primary energy source.

The basic components of a fuel cell are:

— an electrolyte

— an anode

— a cathode.

Typically, other peripherals are also required:

— a reformer

— a desulphuriser

— power conditioning (transformers, inverters)

— a heat management system

— a water management system

— controls and monitoring.

A reformer converts a hydrogen-based fuel (natural gas for example) into hydrogen. Gaseous fuel is continuously fed to the anode while an oxidant (oxygen from the air) is fed continuously to the cathode. The load is connected between the anode and the cathode. Each fuel cell (see Figure 33)

Figure 33:
Simplified diagram of a fuel cell

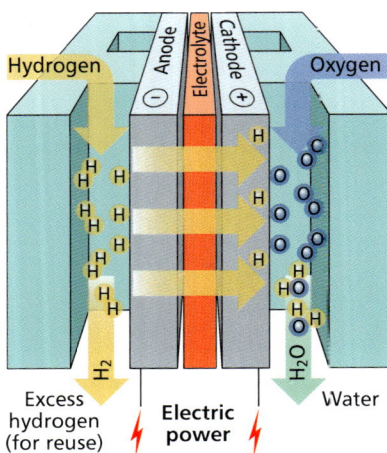

produces in the order of 1 volt so a number of cells are connected together to form a stack to produce a working voltage.

The benefits of a fuel cell are:

— direct conversion (no combustion)

— no moving parts in the fuel cell stack

— quiet

— fuel flexibility

— good part load performance (40–55 per cent efficient)

— waste heat can be used for combined heat and power (CHP)/ combined cooling heat and power (CCHP) improving overall system efficiency.

However, limitations include:

— high cost (at the time of writing)

— unfamiliar technology

— the need for gas on the data centre site

— large footprint per kVA

— high maintenance costs

— short stack life

— sensitive to pollutants

— requires a short time to get up to temperature and take load, limiting its effectiveness as a backup supply

— relatively small ratings commercially available.

There are five basic types of fuel cell:

— proton exchange membrane (PEMFC)

— alkaline (AFC)

— phosphoric acid (PAFC)

— molten carbonate (MCFC)

— solid oxide (SOFC).

At present, only phosphoric acid fuel cells are commercially available for data centre applications.

A specification for fuel cells in the UK is given by BS EN 62282.

14.14   Amorphous metal substation transformers

Transformers incur two types of losses: 'no-load losses' and 'load losses' (no-load losses are present irrespective of loading and load losses are related to the resistance of the windings and are load related).

These no load losses are fundamentally made up of a hysteresis loss (re-orientation of magnetic domains) and eddy current losses (circulating currents within the magnetic material).

While transformers are very efficient, these losses are still significant, accounting for 2–3 per cent of all electricity produced.

No load losses are particularly important for data centre operation, where transformers are typically loaded to less than 40 per cent due to maintenance and redundancy requirements.

Conventional transformers use cold rolled grain-oriented silicon steel (Fe-Si-alloy) however amorphous transformers use an alloy of iron, silicon and boron (Fe-Si-B) to provide a transformer core that has a low hysteresis loss (as it is very easily magnetised and re-magnetised) and has a high resistivity 130 $\mu\Omega$-cm compared with grain oriented electrical steels (CRGO) silicon steel of 51 $\mu\Omega$-cm efficiently reducing eddy current issues. These characteristics reduce no-load losses by up to 70 per cent.

At the time of writing, these transformers are available in ratings of between 100 and 3150 kVA up to 36 kV.

14.15   Direct server cooling

Recent developments, particularly in the USA, have indicated that server manufacturers have attempted to address heat rejection at source using direct server cooling.

These systems incorporate elements and components using complete immersion by adapting 'end-to-end' heat rejection by liquid cooling.

The modules are fully immersed in an inert liquid coolant and rejected to a secondary water-cooled system by means of a heat exchanger installed within the cabinet serving groups of servers.

14.16   IT software development

There are two main issues with IT rack mounted equipment: the inefficiency of the equipment itself and the amount of processing power required.

Semiconductors, by their nature, give off lots of heat and require fans/cooling. However, recent development in super conductors, and more recently laser cooling techniques, are reducing the amount of heat rejected. The focus should always be to co-ordinate with the IT equipment providers to ensure the equipment being installed is as efficient as possible to reflect these ongoing technological advancements. The IT equipment should also be designed with energy-saving modes etc.

Processing power reflects the programmer's topological process. Where in-house programmers are used, programme/code should be as efficient as possible to reduce the amount of processing power required. The reduction in programming code will reduce the processing power and IT equipment, producing an overall reduction in heat generation.

References

ASHRAE (2009) *Design Considerations for Datacom Equipment Centers* (2nd. edn.) (Atlanta, GA: American Society of Heating, Refrigerating and Air-Conditioning Engineers)

BSRIA (2009) *Soft Landings Framework* BG4/2009 (Bracknell: BSRIA)

BSI (1999) BS EN ISO 14644-1: 1999: *Cleanrooms and associated controlled environments. Classification of air cleanliness* (London: British Standards Institution)

BSI (1999–2006) BS EN ISO 14644: *Cleanrooms and associated controlled environments* (London: British Standards Institution)

BSI (2007) BS EN 50173-5: 2007: *Information technology. Generic cabling systems. Data centres* (London: British Standards Institution)

BSI (2009) BS EN 50174: 2009: *Information technology. Cabling installation. Installation specification and quality assurance* (London: British Standards Institution)

BSI (2010) BS 6701: 2010: *Telecommunications equipment and telecommunications cabling* (London: British Standards Institution)

BSI (2011) BS 6266: 2011: *Fire protection for electronic equipment installations. Code of practice* (London: British Standards Institution)

EU (2008) *Code of Conduct on Energy Consumption of Broadband Equipment* (Brussels: European Commission Joint Research Centre) (available at http://re.jrc.ec.europa.eu/energyefficiency/pdf/CoC%20data%20centres%20nov2008/CoC%20DC%20v%201.0%20FINAL.pdf) (accessed July 2012)

TIA (2012) TIA 942: *Telecommunications Infrastructure Standard for Data Centers* (Arlington, VA: Telecommunications Industry Association)

Further reading and weblinks

American Society of Heating, Refrigerating and Air-Conditioning Engineers (ASHRAE) (website) http://www.ashrae.org

DTI (2007) *Meeting the Energy Challenge — A White Paper on Energy* (London: Department for Trade and Industry) (available at http://www.berr.gov.uk/files/file39387.pdf) (accessed July 2012)

Flucker S and Tozer R (2011) 'Data Centre Cooling Air Performance Metrics' *CIBSE Technical Symposium, DeMontfort University, Leicester UK, 6–7 September 2011* (available at http://www.cibse.org/content/cibsesymposium2011/Paper081.pdf)

Intergovernmental Panel on Climate Change (IPCC) (website) http://www.ipcc.ch

Market Transformation Programme (website) http://efficient-products.defra.gov.uk

Telecommunications Industry Association (TIA) (website) http://www.tiaonline.org

The Green Grid (website) http://www.thegreengrid.org

Tozer R (2009) *Global Data Centre Energy Strategy Data Centre Dynamics Notes* (available at http://www.dc-oi.com/publications.htm) (accessed July 2012)

Tozer R, Salim M and Kurkjian C (2009) 'Air management metrics in data centers' *ASHRAE Trans.* **115** (1) 63 (May 2009)

Tozer R, Wilson M and Flucker S (2008) *Cooling challenges for mission critical facilities* (Carshalton: Institute of Refrigeration)

Uptime Institute (website) http://uptimeinstitute.com

Appendix A: Tier classification

Tier classifications are often misunderstood and misquoted. Their original purpose was to enable categorisation of a data centre with respect to functionality, capacity and expected availability (or performance). See Table 4 at the end of this section for a summary of the tier ratings.

This tier classification approach was first introduced by the Uptime Institute (UI) and elements of this initiative have been incorporated into the American National Standards Institute (ANSI) accredited standard TIA 942 (TIA, 2012).

There are basic differences between the UI tier standard and TIA 942. In fundamental terms, the UI standard focuses on power and cooling to critical plant and equipment with respect to maintenance and availability. TIA 942 is a much more prescriptive and wide-ranging document referring to American codes, standards and weather challenges; for this reason, this section focuses on the UI tier standards.

For more information reference should be made to the UI website (see further reading and weblinks).

Table 4:
Summary of tier ratings

	Tier I	Tier II	Tier III	Tier IV
Capacity components To support IT load	N	N	N+1	N After any failure
Concurrently maintainable	No	No	Yes	Yes
Distribution paths	1	1	1 active and 1 alternate	2 simultaneously active
Fault tolerance	No	No	No	Yes
Compartmentalisation	No	No	No	Yes
Requirement for continuous cooling	Load density dependent	Load density dependent	Load density dependent	Required

Appendix B: Abbreviations

ACU: air conditioning unit

ANSI: American National Standards Institute

ASD: aspirating smoke detector

ASHRAE: American Society for Heating, Refrigeration and Air Conditioning Engineers

BCO: British Council for Offices

BMS: building management systems

BREEAM: Building Research Establishment Environmental Assessment Method

BSRIA: Building Services Research and Information Association

CCHP: combined cooling heat and power

CCTV: closed circuit television

CDM: construction design and management

CE: 'Conformité Européenne' – European conformity

CEAS: Certified Environmental Assessment Scheme

CER: central equipment room

CFD: computational fluid dynamics

CHP: combined heat and power

CHW: chilled water

CoMAH: Control of Major Accident Hazards Regulations

COP: coefficient of performance

CRAC: computer room air conditioning

CRAH: computer room air handling

CRGO: grain oriented electrical steels

CSR: corporate social responsibility

DCIM: data-centre infrastructure management

DDA: Disability Discrimination Act

DNO: distribution network operator

DRUPS: diesel rotary uninterruptible power supply

DX: direct expansion

EC: electronically commutated

EDP: electronic data processing

EMI: electro-magnetic interference

EMS: energy management system

EPC: energy performance certificate

EPO: emergency power off

F&R: flow and return

FA: fire alarm

Fabric extender: IT switching device

FAT: factory acceptance test

FM: facilities management

H&S: health and safety

HD: high density

HSSD: high security sampling detectors
HV: high voltage
HVAC: heating, ventilating and air conditioning
IEEE: Institute of Electrical and Electronics Engineers
IET: The Institution of Engineering and Technology
ICP: independent connection provider
iDNO: independent distribution network operator
ISP: internet service provider
IST: integrated system test
IT: information technology
LAN: local area network
LEED: Leadership in Energy and Environmental Design
LV: low voltage
LZC: low and zero carbon
MEP: mechanical electrical and plumbing
MV: medium voltage
NOC: network operation centre
NOX: nitrous oxide
O&M: operation and maintenance
PDU: power distribution unit
PIDS: perimeter intruder detection system
PLC: programmable logic controller
POP: point of presence
PUE: power usage effectiveness
PV: photo voltaic
RFI: request for information
RPP: rack power panel
SCADA: supervisory control and data acquisition
SPOF: single point of failure
STS: static transfer switch
UI: Uptime Institute
UPS: uninterruptible power supply
WAN: wide area network

Index